Nonlinear Electronic Circuits

NONLINEAR ELECTRONIC CIRCUITS

Aldert van der Ziel
University of Minnesota

A WILEY-INTERSCIENCE PUBLICATION

JOHN WILEY & SONS, New York · London · Sydney · Toronto

Library of Congress Cataloging in Publication Data:

Van der Ziel, Aldert, 1910-
Nonlinear electronic circuits.

"A Wiley-Interscience publication."
Bibliography: p.
1. Electronic circuits. 2. Logic circuits.
I. Title.
TK7867.V4 621.3815'3 76-48145
ISBN 0-471-02227-6

Printed in the United States of America

10 9 8 7 6 5 4 3 2 1

Preface

This book provides an introduction to nonlinear circuits with applications to logic circuits. It arose out of a series of lectures given at the University of Minnesota for junior students in electrical engineering who had completed two quarters of electronics.

Chapter 1 introduces some transient responses often encountered in nonlinear circuits. Chapter 2 deals with diode circuits and their transient response. Chapter 3 discusses the transient behavior in nonlinear bipolar transistor circuits, and Chapter 4 treats field-effect transistor circuits in a similar manner. Chapter 5 deals with various types of flip-flop circuits. Chapter 6 gives a brief introduction to Boolean algebra and the electronic implementation of the various logic operations. Chapter 7 applies the results obtained in the previous chapters to logic circuits such as transistor-transistor logic (T^2L), emitter-coupled logic, and drivers. Chapter 8 deals with time bases and blocking oscillators, while Chapter 9 treats generators of sinusoidal waveforms (oscillators). Chapter 10 discusses various flip-flops and latches and their application to the storage and transmission of information. Chapter 11 presents a short introduction to some newer developments.

This book should be helpful in nonlinear electronics courses for junior and beginning senior electrical engineering students. It should also benefit field engineers working with integrated circuits, who are curious about what is going on inside the "black boxes" they are using.

I am indebted to my students for improvements in the manuscript and to my colleagues for help and encouragement. My wife helped to prepare the manuscript; her steady support is greatly appreciated.

Minneapolis, Minnesota
October 1976

A. VAN DER ZIEL

Contents

Nonlinear Electronic Circuits

CHAPTER

1 Pulse Response of Simple *RC* Networks

In discussing the pulse response of electronic circuits, one finds that the response can often be reduced to the response of the two electric circuits shown in Fig. 1.1. For this reason we deal with the pulse response of these circuits in some detail, and we point out how the discussion can be used in later chapters.

In more complicated cases one should solve the transient problem by Laplace transform methods, but in simple circuits the solution is best obtained by solving the differential equation of the system directly.

1.1a Two Simple *RC* Circuits

We first discuss the transient response of the circuit shown in Fig. 1.1*a*. The differential equation of the circuit is

$$C\frac{dV_c}{dt}=\frac{V(t)-V_c(t)}{R} \qquad \text{or} \qquad \frac{dV_c}{dt}+\frac{V_c}{\tau}=\frac{V(t)}{\tau} \tag{1.1}$$

where V_c is the voltage developed across the capacitance C and $\tau=RC$ is the time constant of the network.

We assume that the voltage $V(t)$ has been off for a long time and is turned on at $t=0$. That is, $V(t)=0$ for $t<0$, and $V(t)=V_1$ for $t>0$. We

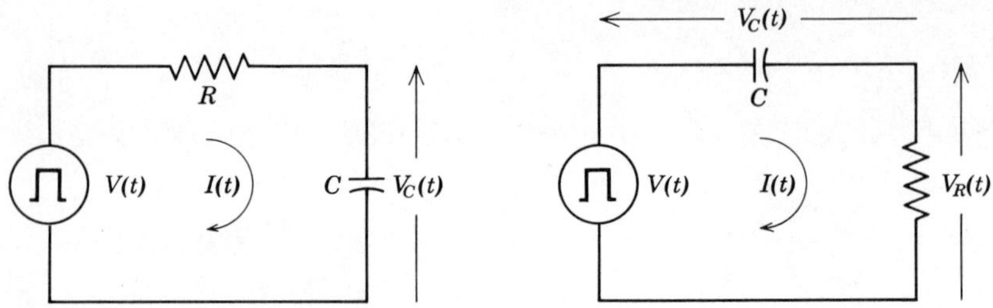

Figure 1.1. (*a*) *RC* network driven by a pulse $V(t)$. (*b*) *CR* network driven by a pulse $V(t)$.

further assume that $V_c(t)=0$ at $t=0$. For $t>0$ Eq. (1.1) can be written

$$\frac{dV_c}{dt}+\frac{V_c}{\tau}=\frac{V_1}{\tau} \tag{1.2}$$

with the initial condition

$$V_c(t)=0 \qquad \text{at } t=0^+ \tag{1.2a}$$

We observe that $V_c=V_1$ is a solution of the inhomogeneous equation, and $A\exp(-t/\tau)$ is the full solution of the homogeneous equation, where A is an arbitrary constant. Therefore the full solution to Eq. (1.2) is

$$V_c(t)=V_1+A\exp\left(-\frac{t}{\tau}\right) \tag{1.3}$$

But we must now satisfy the condition (1.2a). This yields $V_1+A=0$ or $A=-V_1$. Hence the solution

$$V_c(t)=V_1\left[1-\exp\left(-\frac{t}{\tau}\right)\right] \tag{1.4}$$

satisfies both the differential equation and the initial condition. This is called the *turn-on transient* of the circuit in Fig. 1.1*a*. We note that $V_c(\infty)=V_1$. This result could have been obtained directly from Eq. (1.2) by putting $dV_c/dt=0$.

We next assume that the voltage $V(t)$ has been on for a long time and is turned off at $t=0$. That is, $V(t)=V_1$ for $t<0$, and $V(t)=0$ for $t>0$. We further assume that $V_c(t)=V_1$ for $t=0$. For $t>0$ Eq. (1.1) can then be written

$$\frac{dV_c}{dt}+\frac{V_c}{\tau}=0 \tag{1.5}$$

with the initial condition

$$V_c(t) = V_1 \qquad \text{at } t = 0^+ \tag{1.5a}$$

The full solution to Eq. (1.5) is

$$V_c(t) = A \exp\left(-\frac{t}{\tau}\right) \tag{1.6}$$

To satisfy condition (1.5a) we must now require $V_1 = A$, so that

$$V_c(t) = V_1 \exp\left(-\frac{t}{\tau}\right) \tag{1.7}$$

satisfies both the differential equation and the initial condition. This is called the *turn-off transient* of the circuit in Fig. 1.1*a*.

It should be noted that the transient solutions for the stored charge q_s in diodes and transistors are quite similar to the solution for the transient response of the voltage V_c in Fig. 1.1*a*. This is not surprising, for we shall see that the differential equation of one transient problem is transformed into the differential equation of the other problem by substituting q_s for V_c.

We now turn to the transient problem in Fig. 1.1*b*. For the turn-on transient we put $V(t) = 0$ for $t < 0$ and $V(t) = V_1$ for $t > 0$ with $V_c(t) = 0$ at $t = 0^+$; for the turn-off transient we put $V(t) = 0$ and $V_c(t) = V_1$ at $t = 0^+$. The solution is very easily obtained from the solution for Fig. 1.1*a* by observing that the voltage $V_R(t)$ developed across R is given by

$$V_R(t) = V(t) - V_c(t) \tag{1.8}$$

From Eq. (1.4) we thus have for the turn-on transient for $t > 0$,

$$V_R(t) = V_1 \exp\left(-\frac{t}{\tau}\right) \tag{1.9}$$

and from Eq. (1.7) we have for the turn-off transient $t > 0$ [since $V(t) = 0$ for $t = 0^+$],

$$V_R(t) = -V_1 \exp\left(-\frac{t}{\tau}\right) \tag{1.10}$$

We now show that Fig. 1.1*a* represents an *integrating circuit* as long as $|V_c(t)| \ll |V(t)|$, and Fig. 1.1*b* a *differentiating circuit* if $|V_R(t)| \ll |V(t)|$.

The proof of the first statement is as follows. If $|V_c(t)| \ll |V(t)|$, we have

$$I(t) = \frac{dQ_c}{dt} = C\frac{dV_c}{dt} \cong \frac{V(t)}{R}$$

where $Q_c = CV_c$ is the charge stored in the capacitor. Let $V_c(0)=0$. Then we have

$$\frac{dV_c}{dt} = \frac{V(t)}{CR} \qquad \text{or} \qquad V_c(t) = \frac{1}{CR}\int_0^t V(u)\,du$$

as required.

The proof of the second statement is as follows. If $|V_R(t)| \ll |V(t)|$, we have for the charge stored in the capacitor,

$$Q_c \cong CV(t)$$

hence

$$V_R(t) = RI(t) = R\frac{dQ_c}{dt} = RC\frac{dV}{dt}$$

as required.

A somewhat different problem occurs in the pulse response of interstage networks as a result of the coupling capacitor C between the stages. The equivalent circuit is shown in Fig. 1.2; it differs from the one in Fig. 1.1 in that it has two resistors.

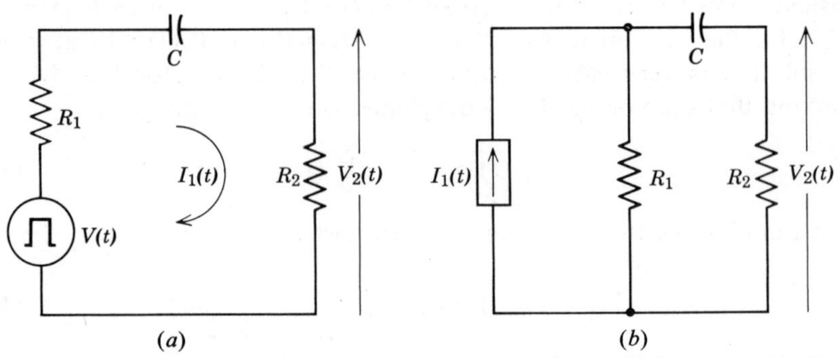

Figure 1.2. (*a*) Interstage network driven by a voltage pulse $V(t)$ in series with R_1. (*b*) Interstage network driven by a current pulse $I(t)$ in parallel with R_1.

We first consider the circuit in Fig. 1.2*a*, with $V(t)=0$ for $t<0$, $V(t)=V_1$ for $t>0$, and $V_c(t)=0$ at $t=0^+$. We then have for $V_c(t)$, from Eq. (1.4),

$$V_c(t) = V_1\left[1-\exp\left(-\frac{t}{\tau}\right)\right] \tag{1.11}$$

where $\tau = C(R_1+R_2)$. The transient current $I_1(t)$ flowing through the circuit is therefore

$$I_1(t)=C\frac{dV_c}{dt}=\frac{C}{\tau}V_1\exp\left(-\frac{t}{\tau}\right) \tag{1.12}$$

so that the voltage developed across R_2 is

$$V_2(t)=I_1(t)R_2=\frac{R_2}{R_1+R_2}V_1\exp\left(-\frac{t}{\tau}\right) \tag{1.13}$$

since $CR_2/\tau=R_2/(R_1+R_2)$ according to the definition of τ.

Usually $V(t)$ comes from a step current generator $I(t)$ in parallel with the resistance R_1 (Fig. 1.2*b*). Then

$$V(t)=I(t)R_1 \tag{1.14}$$

If now $I(t)=0$ for $t<0$ and $I(t)=I_1$ for $t>0$, we have

$$V(t)=0 \qquad \text{for } t<0 \qquad V(t)=I_1R_1 \qquad \text{for } t>0 \tag{1.14a}$$

so that Eq. (1.13) can be written

$$V_2(t)=I_1\frac{R_1R_2}{R_1+R_2}\exp\left(-\frac{t}{\tau}\right)=I_1R_{\text{par}}\exp\left(-\frac{t}{\tau}\right) \tag{1.15}$$

where $R_{\text{par}}=R_1R_2/(R_1+R_2)$ is the resistance of R_1 and R_2 connected in parallel.

In the same way the turn-off transient is evaluated.

1.1b Generalized Solutions

We now return to Fig. 1.1, but generalize the initial conditions. We put

$$V_c(t)=V_c(0^+) \qquad \text{at } t=0^+ \qquad V_c(t)=V_c(\infty) \qquad \text{at } t=\infty \tag{1.16}$$

where $V_c(\infty)$ is obtained from Eq. (1.2) by putting $dV_c/dt=0$. For both the turn-on and the turn-off transient Eq. (1.2) becomes

$$\frac{dV_c}{dt}+\frac{V_c}{\tau}=\frac{V_c(\infty)}{\tau} \tag{1.17}$$

which has the general solution

$$V_c(t)=V_c(\infty)+A\exp\left(-\frac{t}{\tau}\right) \tag{1.18}$$

We must now satisfy the conditions at $t=0^+$. This yields

$$V_c(0^+)=V_c(\infty)+A \quad \text{, or} \quad A=V_c(0^+)-V_c(\infty) \tag{1.18a}$$

so that

$$V_c(t)=V_c(0^+)\exp\left(-\frac{t}{\tau}\right)+V_c(\infty)\left[1-\exp\left(-\frac{t}{\tau}\right)\right] \tag{1.19}$$

which may be written in the alternate form

$$V_c(t)=V_c(\infty)+\left[V_c(0^+)-V_c(\infty)\right]\exp\left(-\frac{t}{\tau}\right) \tag{1.19a}$$

so that $V_c(t)$ goes from $V_c(0^+)$ at $t=0^+$ to $V_c(\infty)$ at $t=\infty$.

In some cases it is inconvenient to refer the transient problem back to $V_c(t)$. It is often much better to refer back to another voltage. For example, in the case of Fig. 1.2 it would be much more convenient to replace Eq. (1.19) by the corresponding equation for $V_2(t)$

$$V_2(t)=V_2(0^+)\exp\left(-\frac{t}{\tau}\right)+V_2(\infty)\left[1-\exp\left(-\frac{t}{\tau}\right)\right] \tag{1.20}$$

We shall encounter several examples of this later in this book. In this particular case $V_2(\infty)=0$.

A similar case occurs in the transient response of diodes and transistors. In the case of transistors we have the following equation for the stored hole charge $q_b(t)$ in the base of an $n-p-n$ transistor:

$$\frac{dq_b}{dt}+\frac{q_b}{\tau}=\frac{q_b(\infty)}{\tau} \tag{1.21}$$

with the initial condition $q_b(t)=q_b(0^+)$ at $t=0^+$. In analogy with Eq. (1.19) the solution is

$$q_b(t)=q_b(0^+)\exp\left(-\frac{t}{\tau}\right)+q_b(\infty)\left[1-\exp\left(-\frac{t}{\tau}\right)\right] \tag{1.22}$$

It now turns out that, for the turn-off transient, $q_b(0^+)$ and $q_b(\infty)$ have opposite signs, with $q_b(0^+)$ positive and $q_b(\infty)$ negative. There is then a time t_0 after which $q_b(t)$ becomes negative. But $q_b(t)$ is by its very nature a positive quantity. This therefore means that the transient stops at $t=t_0$.

Hence

$$0 = q_b(\infty) + \left[q_b(0^+) - q_b(\infty)\right]\exp\left(-\frac{t_0}{\tau}\right)$$

or

$$\exp\left(\frac{t_0}{\tau}\right) = \frac{q_b(0^+) - q_b(\infty)}{-q_b(\infty)}$$

or

$$t_0 = \tau \ln\left[\frac{q_b(0^+) - q_b(\infty)}{-q_b(\infty)}\right] \tag{1.23}$$

PROBLEMS

1.1 In the circuit in Fig. 1.1*a* $V(t)$ is a square wave of 5-V amplitude and 5000-Hz frequency (Fig. 1.3). Plot the steady-state output response $V_c(t)$ for

(a) $RC = 10^{-5}$ s
(b) $RC = 10^{-4}$ s
(c) $RC = 10^{-3}$ s

Hints. 1. By "steady state" we mean the response after the square wave has been on for a long time. 2. Assume a certain value $V_c(t_1)$ after the square wave has just switched sign, follow it for a half cycle, and bear in mind that after half a cycle $V_c(t) = -V_c(t_1)$.

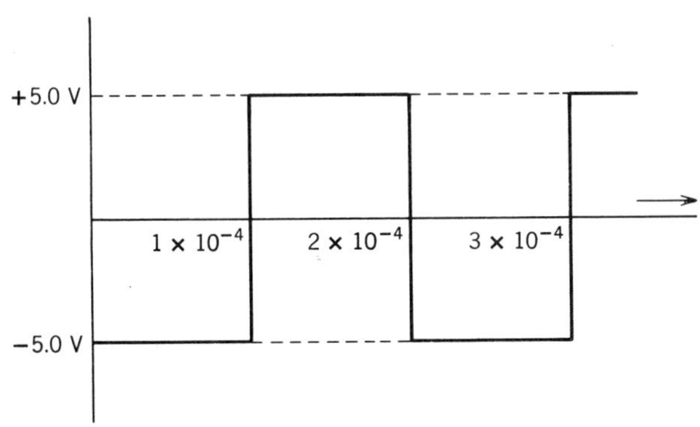

Figure 1.3

1.2 In the circuit in Fig. 1.1*b* $V(t)$ is a square wave of 5-V amplitude and 5000-Hz frequency (Fig. 1.3). Plot the steady-state response $V_R(t)$ for

(a) $RC = 10^{-3}$ s
(b) $RC = 10^{-4}$ s
(c) $RC = 10^{-5}$ s

Hint. Assume a certain value $V_R(t_1)$ after the square wave has just changed sign, follow it for a half cycle until the square wave has switched again, and bear in mind that at that time $V_R(t) = -V_R(t_1)$.

1.3 (a) Find the capacitor C in Fig. 1.2*a* so that $\tau = 0.1$ s if $R_1 = 1000\ \Omega$ and $R_2 = 2000\ \Omega$ (this corresponds to a transistor circuit).

(b) Find the percentage "sag" of $V_2(t)$ in 1 ms.

1.4 Find the capacitor C in Fig. 1.2*a* so that $\tau = 0.1$ s if $R_1 = 1000\ \Omega$ and $R_2 = 10^6\ \Omega$ (this corresponds to an FET circuit).

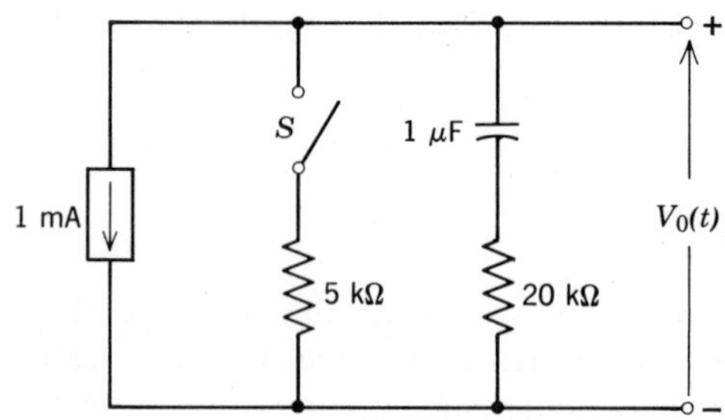

Figure 1.4

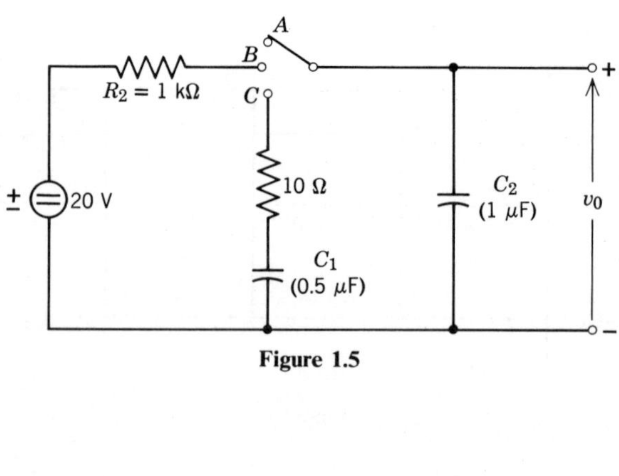

Figure 1.5

1.5 Find the turn-off time t_0 of a transistor if $\tau = 10^{-7}$ s and

(a) $q_b(\infty) = -q_b(0^+)$
(b) $q_b(\infty) = -\frac{1}{2}q_b(0^+)$
(c) $q_b(\infty) = -\frac{1}{5}q_b(0^+)$
(d) $q_b(\infty) = -\frac{1}{20}q_b(0^+)$

and plot t_0 versus $|q_b(\infty)/q_b(0^+)|$.

1.6 The switch S in Fig. 1.4 is closed at all times except for $0 < t < 50$ ms. Calculate, sketch, and label the output waveform of v_0.

1.7 The initial conditions for the circuit in Fig. 1.5 are: Switch is in position A, and charges on C_1 and C_2 are zero for $t < 0$. For $0 < t < 2$ ms the switch is placed in position B; for $t > 2$ ms the switch is placed in position C. Label the waveform of $v_0(t)$.

CHAPTER

2 Diode Circuits

In this chapter we discuss several nonlinear diode circuits. Section 2.1 describes how to determine the operating point of a simple diode circuit and how to represent a nonlinear diode response. Section 2.2 deals with clipper circuits and their application to logic circuits. Section 2.3 considers clamp circuits which shift the dc level of a waveform and clamp the top or bottom of a waveform.

Section 2.4 deals with the transient response of a $p-n$ junction diode, using the charge-control approach when there is charge storage in the diode, and the capacitive approach for the inactive part of the diode response. Section 2.5 applies this knowledge to logic circuits.

2.1 DETERMINATION OF THE OPERATING POINT

Here we deal with finding the operating point of a simple diode circuit. Two principal methods are discussed, the graphical method, also called the load-line method, and the iterative method (Section 2.1b).

In many cases the diode can be represented by an *ideal* diode, that is, a diode for which the diode current $I_D=0$ when the diode voltage $V_D<0$ V, and $I_D=\infty$ for $V_D>0$ V. In other cases a slightly more sophisticated model must be used, consisting of an ideal diode with an emf V_{D0} in series; for silicon diodes $V_{D0}\cong 0.70$ V (Section 2.1a). A piecewise linear model, in which the characteristic has a finite slope near the operating point, turns out to be of little use (Section 2.1c).

2.1a Actual Diodes and Ideal Diodes

A silicon junction diode has a high current for one polarity and a small current for opposite polarity. For that reason it can be idealized by a characteristic of the ideal diode form.

$$I_D = 0 \qquad \text{for } V_D < 0 \qquad I_D = \infty \qquad \text{for } V_D > 0 \tag{2.1}$$

For $V_D = 0$ the current is positive and limited by the external circuit. This characteristic can describe the nonlinear features of most diode circuits. The characteristic of an ideal diode is shown in Fig. 2.1. We shall see how accurate this representation is.

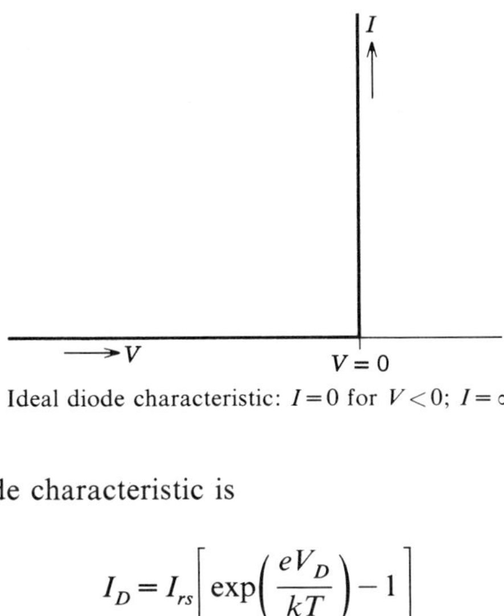

Figure 2.1. Ideal diode characteristic: $I = 0$ for $V < 0$; $I = \infty$ for $V > 0$.

The actual diode characteristic is

$$I_D = I_{rs}\left[\exp\left(\frac{eV_D}{kT}\right) - 1\right] \tag{2.2}$$

where e is the absolute value of the electron charge, k is Boltzmann's constant, and T is the absolute temperature of the device. The current I_{rs} is called the *reverse saturation current* and is of the order of 10^{-15} A at $T = 300°K$; it increases very rapidly with increasing temperature. The parameter kT/e has the dimension of volts and is often represented by the symbol V_T; $V_T = kT/e = 25.8$ mV at $T = 300°K$.

We next ask a few questions which illustrate the meaning of Eq. (2.2).

QUESTION 1. If $I_{rs} = 10^{-15}$ A at $T = 300°K$, at what value of V_D does $I_D = 1$ mA?

Answer. $I_D \gg I_{rs}$ or

$$I_D = I_{rs}\exp\left(\frac{eV_D}{kT}\right)$$

or

$$\exp\left(\frac{eV_D}{kT}\right) = \frac{I_D}{I_{rs}}$$

so

$$V_D = \frac{kT}{e}\ln\left(\frac{I_D}{I_{rs}}\right)$$

or $V_D = 25.8\times10^{-3}\ln 10^{12} = 0.713$ V.

QUESTION 2. If $I_D = I_{D0} = 1$ mA at $V_D = V_{D0} = 0.713$ V at $T = 300°$K, how much must V_D be increased to raise I_D by a factor of 10?

Answer. Put $V_D = V_{D0} + \Delta V_D$. Since $I_{D0} = I_{rs}\exp(eV_{D0}/kT)$,

$$I_D = I_{rs}\exp\frac{e(V_{D0}+\Delta V_D)}{kT} = I_{D0}\exp\left(\frac{e\Delta V_D}{kT}\right) \tag{2.2a}$$

so

$$\exp\left(\frac{e\Delta V_D}{kT}\right) = \frac{I_D}{I_{D0}} = 10 \qquad \Delta V_D = \frac{kT}{e}\ln 10 = 0.0594\ V.$$

This result is independent of the value of the initial current I_{D0}.

In round numbers we can say that, at $T = 300°$K, $V_{D0} \cong 0.70$ V at $I_D = 1$ mA, and that I_D increases by a factor of 10 if V_D increases by 60 mV. The value $V_D = 0.70$ V is called the *turn-on* voltage of the silicon diode.

Since I_D changes so rapidly with increasing V_D, and the current in a diode is usually of the order of 1 mA, an actual silicon diode can be represented by an ideal diode with an emf $V_{D0} = 0.70$ V in series. Figure 2.2*a* shows the equivalent circuit; Fig. 2.2*b* shows the idealized characteristic. If this approximation is not sufficient, higher-order approximations can be easily made (see next Section).

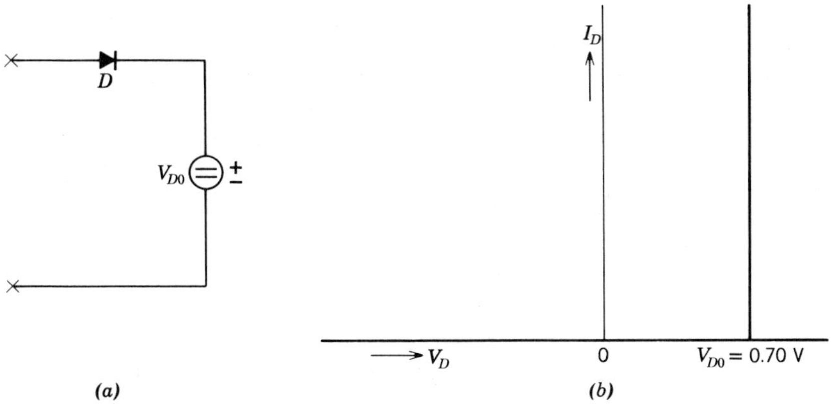

Figure 2.2. (*a*) Equivalent circuit of an actual diode, consisting of an ideal diode and an emf V_{D0} in series. (*b*) Graph representing this characteristic with $V_{D0}=0.70$ V.

2.1b Determination of the Operating Point

There are several ways to determine the operating point (I_D, V_D) for the circuit in Fig. 2.3*a*, consisting of a battery V_{BB}, a resistance R, and a diode D.

The graphical method, illustrated in Fig. 2.3*b* is discussed first. In the first place the current I_D is given by the device characteristic

$$I_D = f(V_D) \tag{2.3}$$

However, I_D is also given by the equation

$$I_D = \frac{V_{BB} - V_D}{R} \tag{2.4}$$

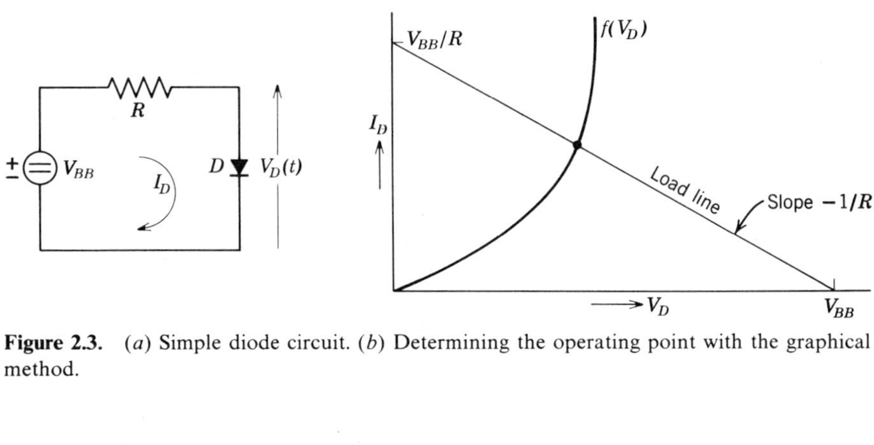

Figure 2.3. (*a*) Simple diode circuit. (*b*) Determining the operating point with the graphical method.

This is a straight line passing through the point $V_D = V_{BB}, I_D = 0$ and having a slope $-1/R$. It is called the *load line*. The operating point (I_D, V_D) is thus obtained as the point of intersection of the characteristic [Eq. (2.3)] and the load line [Eq. (2.4)]; it can be read directly from the graphs. This load-line method of graphical analysis is applicable to any one-port device and can be extended to two-port devices such as field-effect transistors (FETs). The graphical method is nothing but a graphical determination of the root(s) of the equation

$$I_D = f(V_D) = \frac{V_{BB} - V_D}{R} \tag{2.5}$$

We now apply this first to the ideal diode in Fig. 2.1. Then $V_D = 0$ and

$$I_D = \frac{V_{BB}}{R} \tag{2.5a}$$

Next we apply this to the diode model in Fig. 2.2. Then $V_D = 0.70$ V and

$$I_D = \frac{V_{BB} - 0.70}{R} \tag{2.5b}$$

As long as $V_{BB} \gg 0.70$ V, the two approaches give little difference in the diode current I_D. But the two methods differ widely of course in the value of the diode voltage V_D; here the ideal diode model is simply inadequate.

The next method of solution is to evaluate Eq. (2.5) numerically. We use for $f(V_D)$ the actual diode characteristic (2.2) and bear in mind that usually $I_D \gg I_{rs}$; Eq. (2.5) can be written

$$I_D = I_{rs} \exp\left(\frac{eV_D}{kT}\right) = \frac{V_{BB} - V_D}{R} \tag{2.6}$$

We assume that $I_D = I_{D0} = 1.00$ mA at $V_D = V_{D0} = 0.700$ V. Since $I_{D0} = I_{rs} \exp(eV_{D0}/kT)$, Eq. (2.6) can be written

$$I_D = I_{D0} \exp\left(\frac{e\Delta V_D}{kT}\right) = \frac{V_{BB} - V_D}{R} \tag{2.6a}$$

where $\Delta V_D = V_D - V_{D0}$.

This equation can be solved by an iterative method. In this method we start with an initial value $V_D = V_{D0} = 0.700$ V and calculate the current

$$I_{D1} = \frac{V_{BB} - V_{D0}}{R} \tag{2.6b}$$

Next we solve for $\Delta V_D = \Delta V_{D1}$ by solving the equation

$$\frac{I_{D1}}{I_{D0}} = \exp\left(\frac{e\Delta V_{D1}}{kT}\right) \qquad \Delta V_{D1} = \frac{kT}{e}\ln\left(\frac{I_{D1}}{I_{D0}}\right) \tag{2.6c}$$

and evaluate the voltage $V_{D1} = V_{D0} + \Delta V_{D1}$. Then we find the second-order current

$$I_{D2} = \frac{V_{BB} - V_{D1}}{R} \tag{2.6d}$$

and determine $\Delta V_{D2} = V_{D2} - V_{D0}$ from

$$\frac{I_{D2}}{I_{D0}} = \exp\left(\frac{e\Delta V_{D2}}{kT}\right) \qquad \Delta V_{D2} = \frac{kT}{e}\ln\left(\frac{I_{D2}}{I_{D0}}\right) \tag{2.6e}$$

from which $V_{D2} = V_{D0} + \Delta V_{D2}$ can be evaluated. This should be repeated until no further changes in the values result.

We illustrate this method with a few examples.

Example 1. $V_{BB} = 100$ V, $R = 10^4\ \Omega$, and $I_D = 1.00$ mA at $V_D = 0.700$ V.

$$I_{D1} = \frac{V_{BB} - 0.700}{R} = 9.93\ \text{mA}$$

$$\Delta V_{D1} = \frac{kT}{e}\ln\left(\frac{I_{D1}}{I_{D0}}\right) = 25.8\times 10^{-3}\ln 9.93 \cong 0.059\ \text{V}$$

Hence $V_{D1} = V_{D0} + V_{D1} = 0.759$ V. The next approximation gives a negligible change in I_D, hence in V_D, so that the accurate operating point is at $I_D = 9.93$ mA and $V_D = 0.759$ V.

Example 2. $V_{BB} = 10.0$ V, $R = 10^4\ \Omega$, and $I_D = 1.00$ mA at $V_D = 0.700$ V.

$$I_{D1} = \frac{V_{BB} - 0.700}{R} = 0.93\ \text{mA}$$

$$\Delta V_{D1} = \frac{kT}{e}\ln 0.93 = -0.002\ \text{V}$$

so $V_{D1} \cong 0.698$ V in good approximation. No further correction is needed.

Example 3. $V_{BB}=1.00$ V, $R=10^4\ \Omega$, and $I_D=1.00$ mA at $V_D=0.700$ V.

$$I_{D1}=\frac{V_{BB}-0.700}{R}=0.030 \text{ mA}$$

$$\Delta V_{D1}=\frac{kT}{e}\ln\left(\frac{I_{D1}}{I_{D0}}\right)=25.8\times10^{-3}\ln 0.030=-0.090 \text{ V}$$

so

$$V_{D1}=V_{D0}+\Delta V_{D1}=0.610 \text{ V}$$

$$I_{D2}=\frac{V_{BB}-V_{D1}}{R}=0.0390 \text{ mA}$$

$$\Delta V_{D2}=\frac{kT}{e}\ln\left(\frac{I_{D2}}{I_{D0}}\right)=25.8\times10^{-3}\ln 0.0390=-0.084 \text{ V}$$

so $V_{D2}=V_{D0}+\Delta V_{D2}=0.616$ V, which differs from V_{D1} by less than 1%. It may thus be assumed that V_{D2} is reasonably accurate. However, we need one further approximation for I_D. Obviously,

$$I_{D3}=\frac{V_{BB}-V_{D2}}{R}=0.0384 \text{ mA}$$

which differs from I_{D2} by less than 2%. It may thus be assumed that I_{D3} is reasonably accurate. The operating point is therefore at 0.0384 mA and 0.616 V. This example illustrates that the series of approximations converges rapidly and gives accurate results in a few steps.

The conclusion to be drawn from these examples is that the ideal diode approximation gives reasonably accurate values for the diode current I_D as long as V_{BB} is larger than 5–10 V, but that it cannot be used to determine the diode voltage V_D. The approximation method in Fig. 2.2*b* gives reasonably accurate values for the diode current I_D and the diode voltage V_D. The latter holds as long as I_D is of the order of 1 mA; otherwise a simple correction must be made for V_D. If V_{BB} is comparable to 0.70 V, the operating point must be determined by a series of successive approximations, as demonstrated in Example 3.

2.1c Piecewise Linear Approximation

Sometimes a diode characteristic is represented by a piecewise linear characteristic to facilitate the discussion of diode circuits. For the sake of completeness we give this representation, even though rapidly converging approximation methods (Section 2.1b) and computer methods exist that make this representation superfluous.

Let a diode be operating at the point $V_D = V_{D1}, I_D = I_{D1}$. In the vicinity of this operating point the diode then behaves as a resistance $R_d = 1/g_d$, where

$$g_d = \frac{dI_D}{dV_D} = \frac{eI_{rs}}{kT}\exp\left(\frac{eV_{D1}}{kT}\right) = \frac{eI_{D1}}{kT} \tag{2.7}$$

for $I_{D1} \gg I_{rs}$, so

$$R_d = \frac{kT}{eI_{D1}} = \frac{25.8}{I_{D1}}\ \Omega \tag{2.7a}$$

where I_{D1} is in milliamperes. We thus have the piecewise linear characteristic

$$I_D = \frac{V_D - V_{D2}}{R_d} \qquad \text{for } V > V_{D2} \qquad I_D = 0 \text{ otherwise} \tag{2.8}$$

where V_{D2} must be chosen so that the characteristic passes through the point (V_{D1}, I_{D1}). To evaluate V_{D2}, we rewrite Eq. (2.8) as

$$V_D - V_{D2} = I_D R_d \tag{2.8a}$$

and substitute $V_D = V_{D1}$ and $I_D = I_{D1}$. This yields

$$V_{D1} - V_{D2} = I_{D1} R_d = \frac{kT}{e} \qquad \text{or } V_{D2} = V_{D1} - \frac{kT}{e} \tag{2.9}$$

The piecewise linear characteristic is shown in Fig. 2.4.

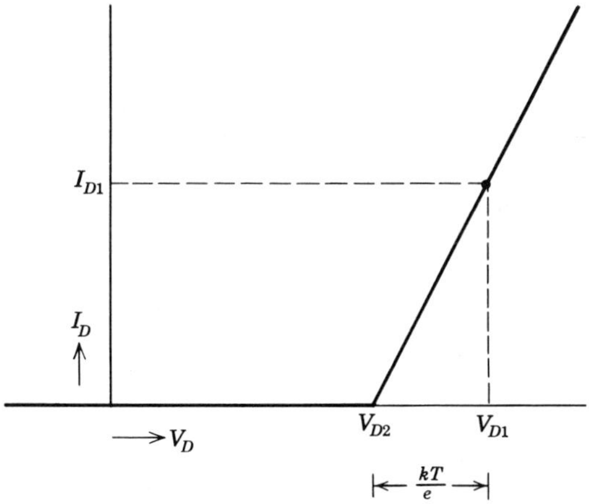

Figure 2.4. Piecewise linear representation of a diode characteristic, valid near the operating point (I_{D1}, V_{D1}).

The disadvantage of the piecewise linear approach is therefore that it is accurate only for a change in V_D of a few millivolts near V_{D0}.

2.2 CLIPPER CIRCUITS

A clipper circuit is a circuit that transmits only parts of a waveform. Since a diode conducts only for *one* polarity, such devices are well suited as clippers. The basic features of these circuits can be understood by treating the diodes as ideal diodes. Afterward one can make corrections for the voltage drop (turn-on voltage) in the actual diode, when necessary.

2.2a Four Basic Clipper Circuits

Figure 2.5 shows four basic clipper circuits and their respective transfer characteristics. They are obtained by changing the polarity of the diode and of the battery.

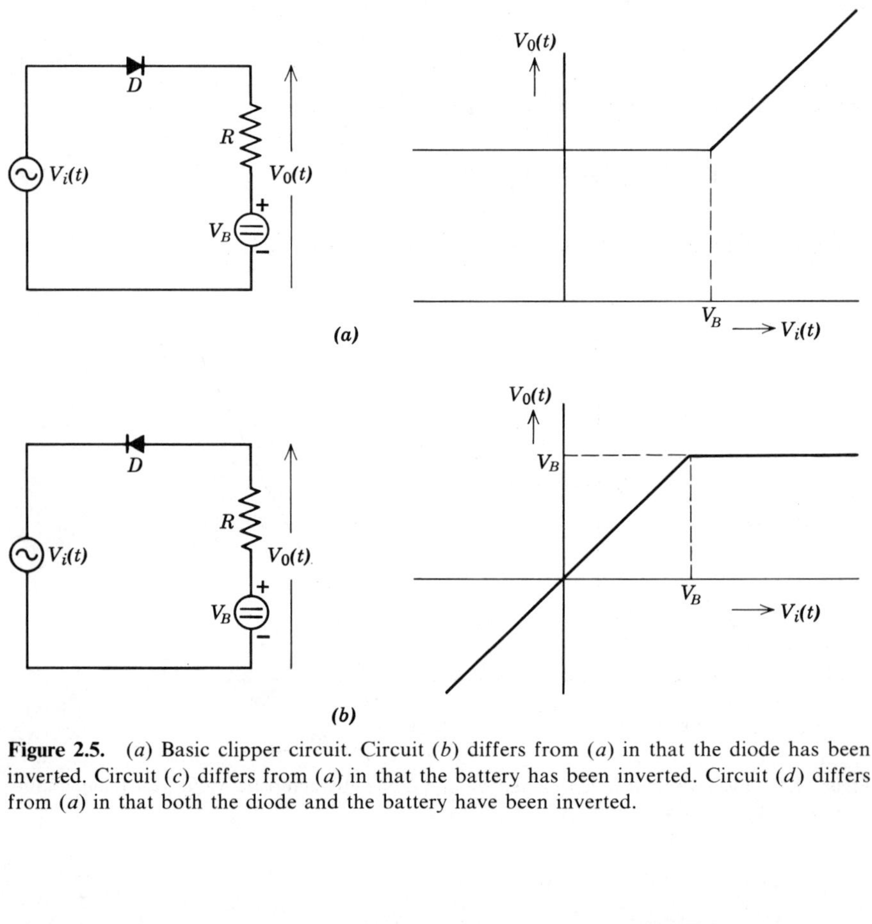

Figure 2.5. (*a*) Basic clipper circuit. Circuit (*b*) differs from (*a*) in that the diode has been inverted. Circuit (*c*) differs from (*a*) in that the battery has been inverted. Circuit (*d*) differs from (*a*) in that both the diode and the battery have been inverted.

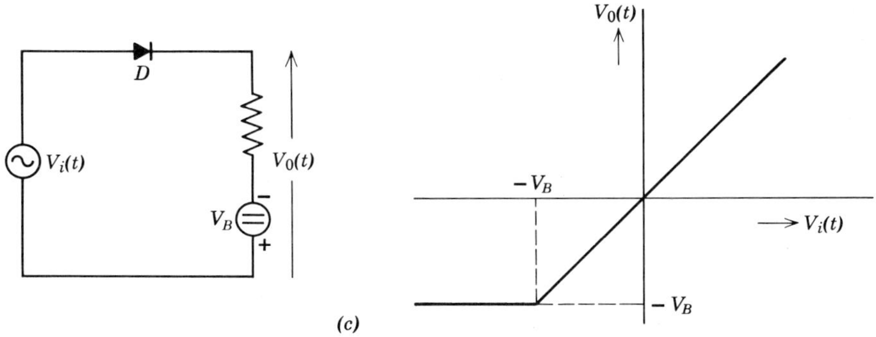

(c)

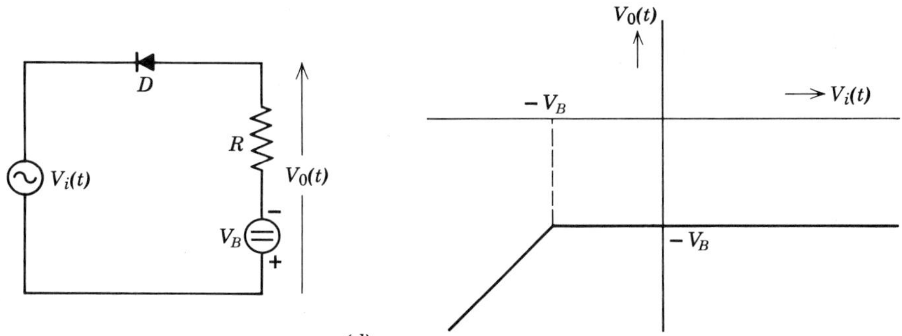

(d)

In Fig. 2.5*a* the ideal diode D does not conduct if $V_i(t) < V_B$, whereas it conducts for $V_i(t) > V_B$. Therefore $V_0(t) = V_B$ for $V_i(t) < V_B$, and $V_0(t) = V_i(t)$ for $V_i(t) > V_B$.

In Fig. 2.5*b* the ideal diode D does not conduct if $V_i(t) > V_B$, whereas it conducts for $V_i(t) < V_B$. Hence $V_0(t) = V_i(t)$ for $V_i(t) < V_B$, and $V_0(t) = V_B$ for $V_i(t) > V_B$.

In Fig. 2.5*c* the ideal D does not conduct if $V_i(t) < -V_B$ and conducts for $V_i(t) > -V_B$. Consequently $V_0(t) = -V_B$ for $V_i(t) < -V_B$, and $V_0(t) = V_i(t)$ for $V_i(t) > -V_B$.

In Fig. 2.5*d* the diode D conducts if $V_i(t) < -V_B$ and does not conduct for $V_i(t) > -V_B$. Therefore $V_0(t) = V_i(t)$ for $V_i(t) < -V_B$, and $V_0(t) = -V_B$ for $V_i(t) > -V_B$.

How are these results altered if we take the voltage drop in the actual diode into account? For the nonconducting part nothing changes: $V_0(t) = V_B$ or $V_0(t) = -V_B$, depending on the polarity of the battery. For the conducting part, however, there is a voltage drop of about 0.70 V in the diode, so $V_0(t) = V_i(t) - 0.70$ V in Fig. 2.5*a* and *c*, and $V_0(t) = V_i(t) + 0.70$ V in Fig. 2.5*b* and *d*.

2.2b Applications

As a first application we take the battery charger in Fig. 2.6. If the diode can be considered ideal, current flows if $V_s(t) > V_B$. Consequently,

$$I(t) = \frac{V_s(t) - V_B}{R + R_s} \qquad \text{for } V_s(t) > V_B \tag{2.10}$$

$$I(t) = 0 \qquad \text{for } V_s(t) < V_B$$

Therefore a short current pulse flows through the battery every cycle. If $V_s(t) = V_{s0}\cos\omega t$, the maximum current flowing through the battery is $I_{\max} = (V_{s0} - V_B)/(R + R_s)$ and the maximum back bias developed across the diode is $-(V_{s0} + V_B)$. The diode must thus be able to withstand this forward current and this back bias.

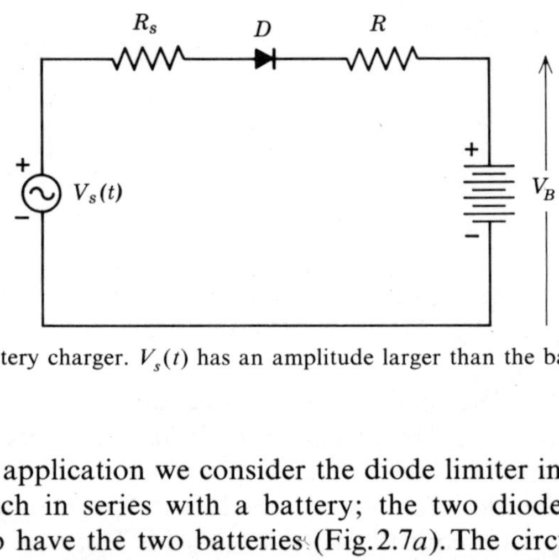

Figure 2.6. Battery charger. $V_s(t)$ has an amplitude larger than the battery voltage V_B.

As a second application we consider the diode limiter in Fig. 2.7. It uses two diodes, each in series with a battery; the two diodes have opposite polarity and so have the two batteries (Fig. 2.7*a*). The circuit transmits for $-V_B < V_s(t) < V_B$, whereas $V_0(t) = -V_B$ for $V_s(t) < -V_B$, and $V_0(t) = V_B$ for $V_s(t) > V_B$. The transfer characteristic is shown in Fig. 2.7*b*. If $V_s(t) = V_{s0}\cos\omega t$ and $V_{s0} > V_B$, the circuit transforms a sinusoidal waveform into a trapezoidal one.

If the two voltages are different, one can transmit a different part of the waveform. Figure 2.8*a* shows a pulse that has noise associated with the base line and with the top. By choosing a circuit that passes the waveform for $V_{B1} < V_s(t) < V_{B2}$, one can completely eliminate the noise on the waveform. Figure 2.8*b* shows the circuit that performs this function.

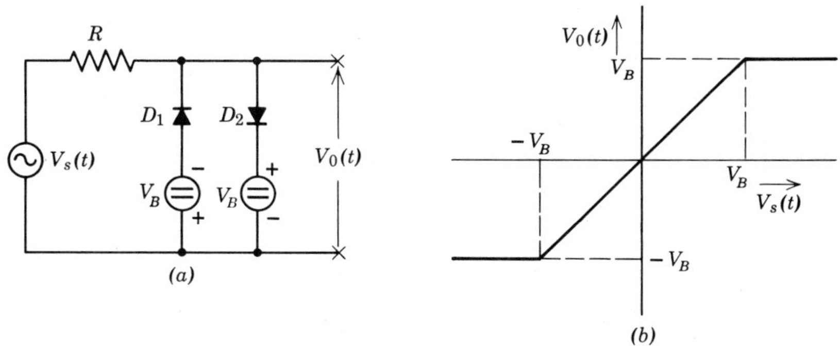

Figure 2.7. Symmetrical diode limiter. (*a*) The circuit. (*b*) The transfer characteristic.

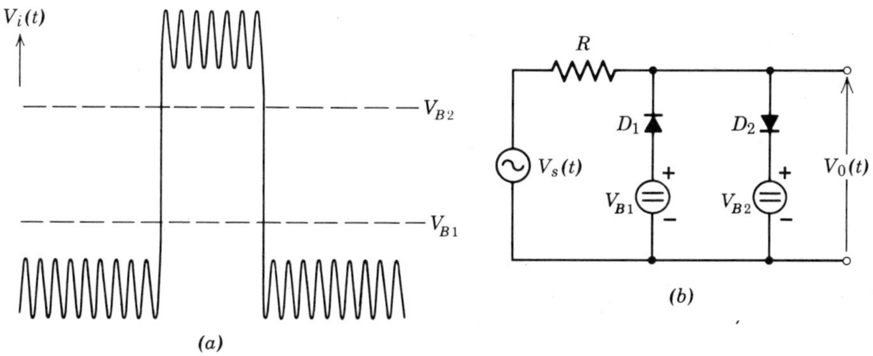

Figure 2.8. (*a*) Waveform with noise on it. (*b*) Asymmetrical limiter circuit for clipping the noise.

A more complicated diode circuit is shown in Fig. 2.9. Three regions of the input and transfer characteristic must be considered:

1. $V_1 < 10$ V. In this case diode D_2 conducts, and diode D_1 is off. This keeps V_A at 10 V, so

$$I_1 = 0 \qquad V_0 = 10 \times \frac{10^4}{10^4 + 10^4} = 5 \text{ V}$$

and the current flowing through the output is $I_0 = 0.5$ mA; I_0 keeps this value as long as V_A stays at 10 V.

2. $10 < V_1 < 15$ V. In this case D_2 and D_1 both conduct, $V_A = 10$ V, and

$$I_1 = (V_1 - V_A) \times 10^{-4} = (V_1 - 10) \times 10^{-4} \text{ A}$$

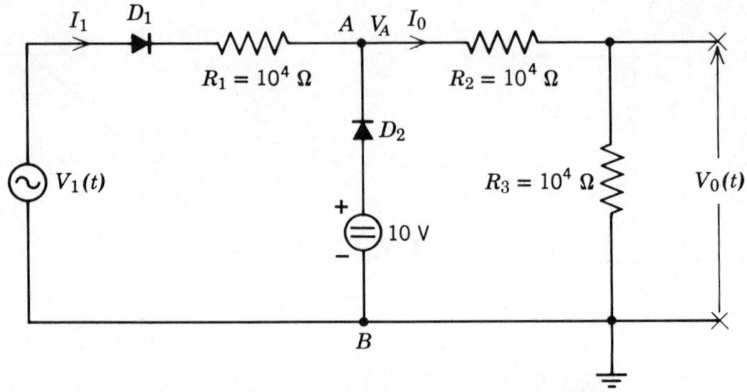

Figure 2.9. More complicated diode circuit, from which the input response and the signal transfer response are computed.

so $I_{D2} = I_0 - I_1 = 0.5\text{ mA} - I_1$ as long as V_A stays at 10 V. For $V_1 = 10$ V all output current I_0 passes through D_2, whereas $I_1 = 0.5$ mA for $V_1 = 15$ V; hence $I_{D2} = 0$, so that all of I_0 comes from D_1. The limits of this range are thus clearly established, and the output voltage V_0 stays at 5 V.

3. $V_1 > 15$ V. In this case $V_A > 10$ V, and D_2 does not conduct. Consequently,

$$I_1 = \frac{V_1}{3\times 10^4} = 0.333\,V_1 \times 10^{-4}\text{ A}; \qquad V_0 = \frac{1}{3}V_1$$

The input and transfer characteristics are therefore:

$V_1 < 10$ V	$I_1 = 0$ and $V_0 = 5$ V
$10\text{ V} < V_1 < 15$ V	$I_1 = (V_1 - 10)\times 10^{-4}$ A and $V_0 = 5$ V
$V_1 > 15$ V	$I_1 = 0.333\,V_1 \times 10^{-4}$ A and $V_0 = \frac{1}{3}V_1$

The input characteristic thus has two breakpoints, one at $V_1 = 10$ V and one at $V_1 = 15$ V, whereas the transfer characteristic has a breakpoint only at 15 V.

The method used here is generally applicable. One always looks for branches between nodal points that contain a diode, for example the branch between points A and B containing the diode D_1 and the emf V_1, and observes that the diode does not conduct if $V_1 < V_{AB}$, whereas the diode conducts for $V_1 > V_{AB}$, and that the current in the branch is then $I_1 = (V_1 - V_{AB})/R$. Here V_{AB} is the voltage between the points A and B, and R is the resistance in the branch.

2.2c Diode Logic Circuits

In digital computers binary arithmetic is used. Closely related to this kind of arithmetic is *Boolean algebra*, which is an algebra of the numbers 0 and 1. We shall see that electronic circuits can perform this kind of algebra.

In Boolean addition $(A+B)$ one has the following rules

$$0+0=0$$

$$1+0=0+1=1$$

$$1+1=1$$

In Boolean multiplication $(A\times B=AB)$ the rules are

$$0\times 0=0$$

$$1\times 0=0\times 1=0$$

$$1\times 1=1$$

We further need the operation $\bar{A}$ (A not) which changes a 0 into a 1 and a 1 into a 0. This is called *inversion*.

$$\bar{0}=1 \qquad \bar{1}=0$$

Assigning the words "yes" and "no" to 1 and 0, respectively, the above operations become *logic* operations.

We now see how diode circuits can be used to perform the first two operations. First one has to define 0 and 1 in terms of voltages. In *positive logic*, 1 is defined as a high voltage V_1, say $V_1>2.0$ V, and 0 is defined as a low voltage V_0, say $V_0<0.5$ V. We shall see that addition and multiplication can be performed with the help of clipper circuits.

Addition is performed by the circuit in Fig. 2.10*a*, which is called the *or* circuit. If there is no voltage in either channel, $V_0=0$ V. If there is an input voltage $V_i(t)$ in *either one* of the channels, the output voltage $V_0(t)=V_i(t)-0.70$ V, which is a 1, so the or circuit indeed performs Boolean addition.

The *and* circuit in Fig. 2.10*b* performs Boolean multiplication. This circuit gives 0 output unless there is a voltage $V_i(t)$ in each channel *simultaneously*. For example, if $V_i(t)\geqslant V_B$ in both channels simultaneously, diodes D_1 and D_2 are turned off and $V_0(t)=(V_B-0.70)R_2/(R_1+R_2)$ which is a 1. If there is no voltage $V_i(t)$ in at least one channel, $V_A=0.70\,V$ and $V_0(t)\cong 0$ V, and this is a 0. The and circuit thus performs multiplication; the third diode is introduced to make $V_0(t)\cong 0$ V if $V_i(t)=0$ V.

Transistor and FET circuits can be so designed that they transform a 0 into a 1 and a 1 into a 0. They thus act as *inverters*.

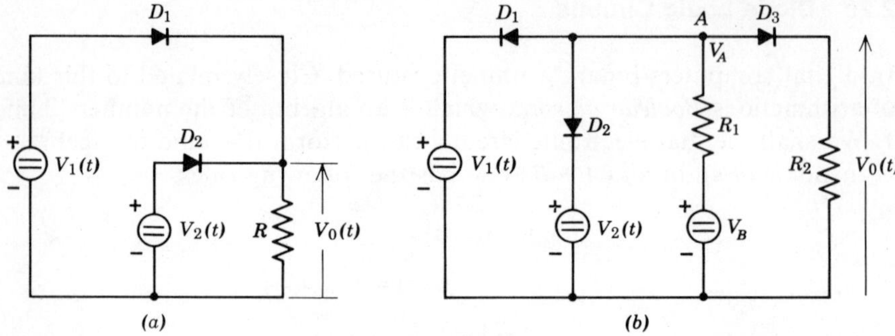

Figure 2.10. (*a*) An or circuit, performing Boolean addition. (*b*) An and circuit, performing Boolean multiplication.

Therefore these two diode circuits and the inverter circuit perform all Boolean operations. All binary operations can be performed by proper combinations of these circuits. We come back to this later.

We note that in each diode circuit the 1 output is somewhat smaller than the 1 input. It is therefore not possible to put many of these circuits one behind the other. By using a combination of diode and transistor circuits this difficulty can be overcome.

Thus far we have discussed *positive* logic, that is, logic in which a high voltage represents a 1 and a low voltage represents a 0. Sometimes one uses *negative* logic, that is, logic in which a high voltage represents a 0 and a low voltage represents a 1.

Since in Fig. 2.10*a* a high voltage in either one of the channels corresponds to a high voltage at the output and a low voltage output results only if $V_1(t)$ and $V_2(t)$ have zero or negative values, Fig. 2.10*a* represents an and circuit for negative logic. In the same manner it can be seen that the circuit in Fig. 2.10*b* represents an or circuit for negative logic.

2.3 CLAMP CIRCUITS

A clamp circuit is a circuit that shifts the dc level of a waveform and clamps either the top or the bottom of a waveform to a fixed bias. We shall see that a combination of a diode and a capacitor in series can perform this operation.

The two basic configurations of the clamp circuit are shown in Fig. 2.11*a* and *b*; they differ in the polarity of the diode. To operate satisfactorily, the diode must be bypassed by a resistor R such that the RC time constant is somewhat larger than the period of the input signal. If the RC time constant is too small, the capacitor C discharges notably during the

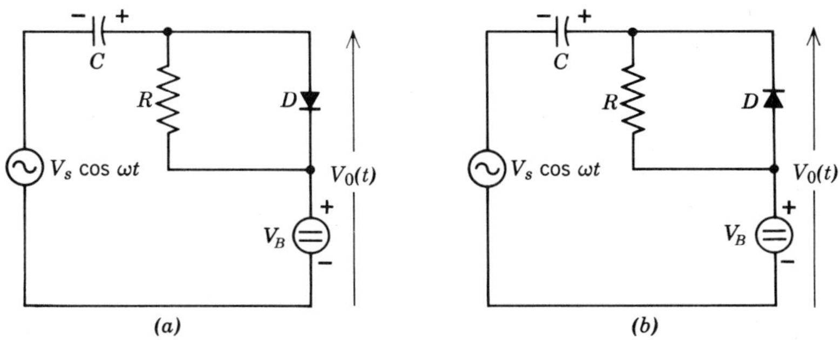

Figure 2.11. Two basic clamp circuits. Circuit (*b*) differs from (*a*) in that it has the diode inverted. Circuit (*a*) clamps the top and (*b*) the bottom of a waveform.

period of the waveform; if the *RC* time constant is much too large, the output voltage cannot follow slow variations in the input waveform. If the resistor *R* is omitted, the circuit either clamps permanently or not at all.

To understand the operation of these circuits, we first turn to Fig. 2.11*a*. If $V_s = 0$, the capacitor charges to V_B, with the polarity as shown in the figure. If the input signal is turned on, so much charge flows *out of* the capacitor *C* that further discharging is prevented, which means that the voltage across *C* will be $-V_s + V_B$. So the voltage developed across the diode *D* is $-V_s + V_s \cos \omega t$, which is always negative except for $\omega t = 0, 2\pi, \ldots$. The output voltage $V_0(t)$ is therefore

$$V_0(t) = V_s \cos \omega t + (-V_s + V_B) = V_B + V_s(-1 + \cos \omega t) \qquad (2.11)$$

This expresses the fact that the *top* of the waveform is clamped to V_B.

We now turn to the circuit in Fig. 2.11*b*. If $V_s = 0$, the capacitor charges to V_B with the polarity shown. If the input signal is now turned on, so much charge flows *into* the capacitor *C* that further charging is prevented. This is the case when the voltage across *C* is $V_s + V_B$, so the voltage developed across the diode *D* is $-V_s - V_s \cos \omega t$, which is always negative except for $\omega t = \pi, 3\pi, \ldots$. The output voltage $V_0(t)$ is therefore

$$V_0(t) = V_s \cos \omega t + (V_s + V_B) = V_B + V_s(1 + \cos \omega t) \qquad (2.11a)$$

This indicates that the *bottom* of the waveform is clamped to V_B.

It should be noted that the clamping of either the top or the bottom of the waveform depends on the *polarity* of the diode; the polarity of the battery determines whether the waveform is clamped to a positive or a negative voltage.

Sometimes the resistor R does not bypass the diode but is connected as shown in Fig. 2.12. We shall see that this diode does not clamp at all as long as $V_s < V_B$. If V_s is first zero, the diode D is back-biased and $V_0(t)=0$ V, so the capacitor C is not charged. If the input voltage is turned on and $V_s < V_B$, the voltage across the diode becomes $-V_B + V_s \cos \omega t$, which is always negative, so the diode D never conducts. The capacitor C never charges, and the circuit does not clamp. The output voltage thus follows the input waveform.

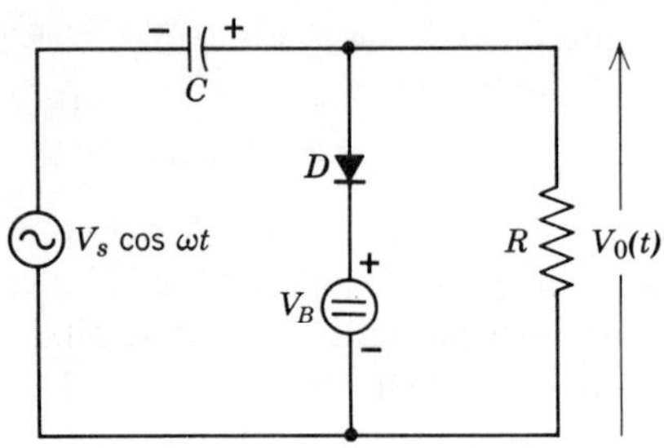

Figure 2.12. A clamp circuit that only works if $V_s > V_B$.

What happens if $V_s > V_B$? In this case the capacitor C is charged to the voltage $-V_s + V_B$ with the polarity shown, hence

$$V_0(t) = V_s \cos \omega t + (-V_s + V_B) = V_B + V_s(-1 + \cos \omega t) \quad (2.11b)$$

which corresponds to Eq. (2.11). The top of the waveform is therefore clamped to V_B.

The difficulty arising when $V_s < V_B$ is avoided if the resistor R is connected across D.

2.4 PULSE RESPONSE OF THE p–n JUNCTION DIODE

Here we discuss the pulse response of p–n junctions. First we consider small-signal admittance of the diode at higher frequencies, taking into account the effects of the stored charge (Section 2.4a). Then we deal with the transient response of the junction, using the charge-control approach (Section 2.4b); this covers the "active" part of the transient response, when carrier injection is important. Section 4.2c discusses the initial and final phases of the transient response, when carrier injection is negligible and the diode acts as a small capacitor. Next, methods of speeding up the transient response are discussed (Section 2.4d). Then a small refinement in the charge-control method is introduced, which speeds up the transients by a factor of about 2 (Section 2.4e).

2.4a Small-Signal Admittance of the Diode at High Frequencies

At higher frequencies the small-signal impedance of the diode becomes complex. It is found that the equivalent circuit can be approximately represented by a small-signal resistance R_d (Fig. 2.13) in parallel with a small-signal capacitance C_d. Here

$$R_d = \frac{1}{g_d} \qquad g_d = \frac{dI_D}{dV_D} = \frac{eI_{D0}}{kT} \qquad \text{or} \qquad R_d = \frac{kT}{eI_{D0}} \tag{2.12}$$

as before. The capacitance C_d is usually much larger than the small-signal capacitance C_j of the space-charge region; it is caused by the storage of holes in the n-region in a p^+–n diode (or by the storage of electrons in the p-region in an n^+–p diode).

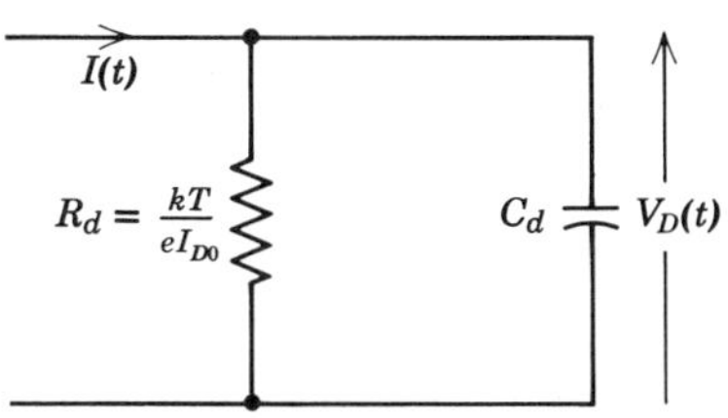

Figure 2.13. Equivalent circuit of a diode, consisting of a parallel RC-circuit representation.

This can be understood as follows.* The holes injected into the n-region of a $p^+ - n$ diode are removed by recombination with electrons. One can therefore introduce a lifetime τ for the injected holes; as a consequence the charge stored in the n-region is finite. Let this charge be denoted by Q_s; then the steady-state recombination current will be Q_s/τ. However, since all the injected holes will recombine eventually, this must be equal to the steady-state current I_D, so

$$I_D = \frac{Q_s}{\tau} \qquad \text{or} \qquad Q_s = I_D\tau \tag{2.13}$$

The small-signal storage capacitance C_s is given by

$$C_s = \frac{dQ_s}{dV_D} = \tau\frac{dI_D}{dV_D} = \frac{\tau}{R_d} \qquad \text{or} \qquad C_sR_d = \tau \tag{2.13a}$$

*In a $p^+ - n$ diode practically all the current is carried by holes; the term $I_{rs}\exp(eV_D/kT)$ in Eq. (2.1) is then due to holes *injected* from the p-region into the n-region.

The product C_sR_d is thus independent of the current and equal to the carrier lifetime τ.

If we were to equate C_s to C_d, we would have

$$C_dR_d = \tau \tag{2.13b}$$

independent of the current. But this result is incorrect; C_d is about $\frac{1}{2}C_s$, for the following reason. A calculation shows that the input admittance Y of a diode can be written as*

$$Y = \frac{dI_D}{dV_D}(1+j\omega\tau)^{1/2} \cong \frac{dI_D}{dV_D}\left(1+\frac{1}{2}j\omega\tau\right)$$

$$= \frac{1}{R_d} + \frac{1}{2}j\omega\frac{\tau}{R_d} = \frac{1}{R_d} + \frac{1}{2}j\omega C_s = \frac{1}{R_d} + j\omega C_d \tag{2.13c}$$

since $(1+j\omega\tau)^{1/2} \cong 1 + \frac{1}{2}j\omega\tau$ if $\omega\tau$ is not too large. Consequently $C_d = \frac{1}{2}C_s$, as stated.

2.4b Transient Behavior of the p–n Junction Diode

Figure 2.14*a* shows a circuit for the testing of the pulse response of diodes. At $t=0$ the voltage $V(t)$ is raised from zero to V_1; we assume for the sake of simplicity that $V_1 > 0.70$ V, the turn-on voltage of the diode. It is then found that $V_D(t)$ rises very rapidly, in a time less than 0.1τ, to the final value, which is of the order of 0.70 V. This is the turn-on transient (Fig. 2.14*b*).

However, when the circuit has been on for a long time with $V(t)$ at V_1, and $V(t)$ is turned off at $t=0$, the output voltage $V_D(t)$ drops gradually from about 0.70 V to 0 V in a time comparable to τ (Fig. 2.14*c*). This is the turn-off transient. These observations require an explanation.

If we try to write a differential equation for $V_D(t)$, we run into considerable difficulty, since the diode characteristic is so strongly nonlinear. We must therefore search for a variable that changes the problem into a linear one. We saw in Section 2.4a that the stored hole charge Q_s in a p^+–n diode varies linearly with the current I_D. It thus seems worthwhile to investigate whether or not $Q_s(t)$ is a suitable variable.

If a current $I(t)$ is passed through a p^+–n diode, part (dQ_s/dt) of this current is used to provide for the buildup of the stored charge $Q_s(t)$, and part (Q_s/τ) gives the approximate recombination current. The time depen-

*Compare, for example, A. van der Ziel, *Solid State Physical Electronics*, 3rd ed. Prentice Hall, Englewood Cliffs, N.J., 1976, Chapter 15.

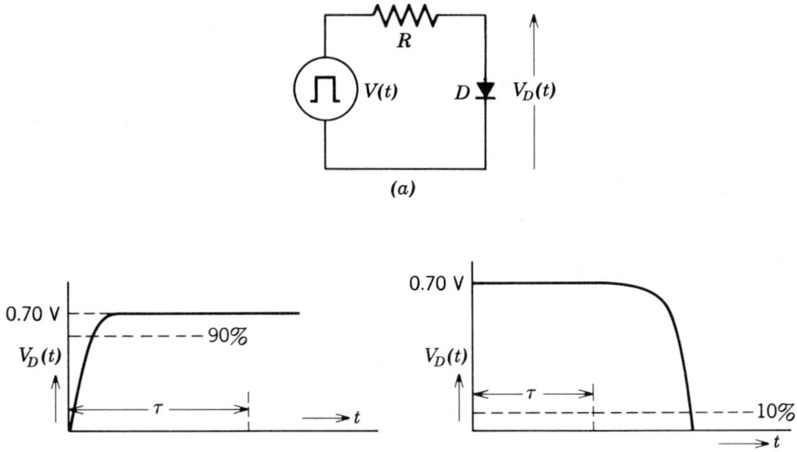

Figure 2.14. (*a*) Simple diode pulse circuit. (*b*) Illustration of the very fast turn-on response. (*c*) Illustration of the rather slow turn-off response.

dence of the stored charge is thus given by the differential equation

$$\frac{dQ_s}{dt} + \frac{Q_s}{\tau} = I(t) \tag{2.14}$$

which is a simple extension of Eq. (2.13). The model introduced here is called the *charge-control model* of the diode. We see that the differential equation is linear, as required. Also, since τ is independent of operating conditions, the equation can be used for large-signal transients. At the end we must transform back from the stored charge to the diode voltage.

It may be shown that Eq. (2.14) corresponds to putting $C_d R_d = \tau$, so that it is not fully correct. Qualitatively, however, the transient behavior is well described by Eq. (2.14). The inaccuracy introduced here is approximately corrected in Section 2.4e.

We now solve this equation for the turn-on transient. Since the circuit is turned on at $t=0$, we have as the initial condition: $Q_s(t)=0$ at $t=0$. Next we evaluate $I(t)$.

$$I(t) = \frac{V_1 - V_D(t)}{R} \cong \frac{V_1 - 0.70}{R} = I_1 \qquad \text{for } t>0 \tag{2.15}$$

since $V_D(t) \cong 0.70$ V most of the time. It is thus seen by inspection that the solution to Eq. (2.14) under the above initial condition can be written as

$$Q_s(t) = I_1\tau\left[1 - \exp\left(-\frac{t}{\tau}\right)\right] \tag{2.16}$$

since this solution satisfies the initial condition as well as the differential equation. $Q_s(t)$ thus rises gradually from the value $Q_s(0) = 0$ to the value $Q_s(\infty) = I_1\tau$ for $t \gg \tau$.

We now solve for the turn-off transient. We assume that $V(t) = V_1$ has been on for $t < 0$ and is turned to $-V_2$ at $t = 0$. Then the initial condition is

$$Q_s(t) = I_1\tau \qquad \text{for } t = 0^+ \tag{2.17}$$

As long as there is still hole charge stored in the n-region, $V_D(t) \cong 0.70$ V, and a current

$$I(t) = \frac{-V_2 - V_D(t)}{R} \cong \frac{-V_2 - 0.70}{R} = -I_2 \tag{2.18}$$

is flowing. Therefore the differential equation of the turn-off transient is

$$\frac{dQ_s}{dt} + \frac{Q_s}{\tau} = -I_2 \tag{2.19}$$

for $t > 0$, as long as $Q_s(t) > 0$. The initial condition is given by Eq. (2.17). As seen by inspection, the solution is, since $Q_s(\infty) = -I_2\tau$,

$$Q_s(t) = I_1\tau\exp\left(-\frac{t}{\tau}\right) - I_2\tau\left[1 - \exp\left(-\frac{t}{\tau}\right)\right] \tag{2.20}$$

as long as $Q_s(t) > 0$. This part of the transient stops at t_1, where $Q_s(t_1) = 0$, or

$$(I_1 + I_2)\tau\exp\left(-\frac{t_1}{\tau}\right) - I_2\tau = 0 \tag{2.21}$$

so

$$\exp\left(\frac{t_1}{\tau}\right) = \frac{I_1 + I_2}{I_2} \qquad t_1 = \tau\ln\left(\frac{I_1 + I_2}{I_2}\right) \tag{2.22}$$

If I_2 is of the order of I_1, the turn-off time is of the order of τ, since then $t_1 \cong \tau\ln 2 = 0.69\tau$. Usually $V_2 = 0$; then I_2 is considerably smaller than I_1, and the turn-off time is longer.

We now transform back from the stored charge $Q_s(t)$ to the diode voltage $V_D(t)$. Knowing the stored charge $Q_s(t)$, we can evaluate the diode voltage $V_D(t)$ as follows. Since $I_D = I_{rs}[\exp(eV_D/kT) - 1]$, we have

$$Q_s = Q_{rs}\left[\exp\left(\frac{eV_D}{kT}\right) - 1\right] \qquad \text{or } \exp\left(\frac{eV_D}{kT}\right) = \frac{Q_s}{Q_{rs}} + 1 \tag{2.23}$$

or

$$V_D(t) = \frac{kT}{e}\ln\left[\frac{Q_s(t)}{Q_{rs}} + 1\right] \tag{2.23a}$$

First we consider the turn-on transient. Substituting Eq. (2.16) into Eq. (2.23a) and assuming $Q_s/Q_{rs} \gg 1$,

$$V_D(t) \cong \frac{kT}{e}\left\{\ln\left(\frac{I_1}{I_{rs}}\right) + \ln\left[1 - \exp\left(-\frac{t}{\tau}\right)\right]\right\} \tag{2.24}$$

We now observe that $(kT/e)\ln(I_1/I_{rs})$ is of the order of 0.70 V, and that $(kT/e)\ln[1 - \exp(-t/\tau)]$ is of the order of -0.06 for $t/\tau = 0.10$. Therefore the diode voltage $V_D(t)$ rises very quickly to 90% of the final value in a time of the order of 0.10 τ, in agreement with Fig. 2.14*b*.

Next we discuss the turn-off transient. Since $Q_s(t) \gg Q_{rs}$ most of the time, little error is made in putting $Q_s(t) = 0$ at the end of the transient. Therefore, in very good approximation, the turn-off time is given by Eq. (2.22), in agreement with Fig. 2.14*c*.

2.4c Effect of the Capacitance C_j of the Space-Charge Region: Turn-on Delay

In the above calculation we made a small error which must now be corrected. The error is due to the fact that the storage capacitance C_s is negligible as long as little current flows, that is, for $V_D < 0.55$ V. When this is the case $C_d \cong C_j$, where C_j is the small-signal capacitance of the space-charge region of the p–n junction. The equivalent circuit is then as shown as in Fig. 2.15. Conversely, when the diode is turned off $[V(t) = -V_2]$, the storage capacitance is already negligible for $V_D < 0.55$ V, and the discharging of the capacitance C_j is again governed by the circuit in Fig. 2.15. We now have for the charging of C_j that $V(t) = V_1$ for $t > 0$, and

$$V_D(t) = V_1\left[1 - \exp\left(-\frac{t}{\tau_j}\right)\right] \tag{2.25}$$

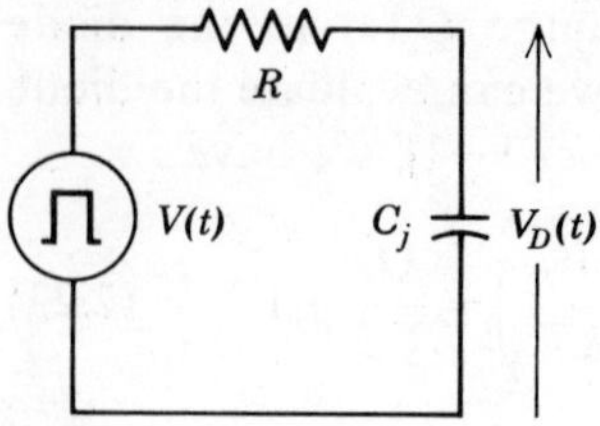

Figure 2.15. Diode circuit for the case in which the diode draws negligible or small currents.

where $\tau_j = C_j R$. This part of the transient stops if $V_D(t) \cong 0.55$ V, and the solution in Section 2.4b takes over. This is called the *turn-on delay*.

Example 1. If $V_1 = 5.0$ V, $C_j = 0.5$ pF, and $R = 2000\ \Omega$, the duration t_1 of the charging of C_j follows from $\tau_j = 0.5 \times 10^{-12} \times 2000 = 10^{-9}$ s.

$$0.55 = 5.0\left[1 - \exp\left(-\frac{t_1}{\tau_j}\right)\right] \cong 5.0\frac{t_1}{\tau_j} \qquad t_1 \cong 0.11\tau_j = 1.1 \times 10^{-10}\ \text{s}$$

For the discharging of C_j we have $V_D(0^+) = 0.55$ V and $V_D(\infty) = -V_2$, hence

$$V_D(t) = 0.55\exp\left(-\frac{t}{\tau_j}\right) - V_2\left[1 - \exp\left(-\frac{t}{\tau_j}\right)\right] \qquad (2.25a)$$

This reaches 0.90 of the final value for $t \cong 2.3\tau_j$. This is called the *recovery time*.

The solution in Section 2.4b must thus be corrected in this manner.

2.4d Speedup of the Response

There are several ways to speed up the response. One way consists of reducing the lifetime τ of the injected carriers. Diodes with a very short lifetime are commercially available or can be made by connecting a transistor as a diode.

The latter statement requires an explanation. The best diode connection for a transistor is to connect the base and the collector. The emitter then injects carriers into the base, and these are almost immediately collected by the collector. The time constant involved in this case is the travel time τ_d of the carriers through the base region. Calculation shows that for a

p–n–p transistor this can be expressed as (see Section 3.4)

$$\tau = \tau_d = \frac{w^2}{2D_p} \tag{2.26}$$

where w is the width of the base region and D_p is the diffusion constant of the holes in the base region.

Example 2. Find τ_d for a silicon p–n–p transistor connected as a diode if $w = -10^{-4}$ cm and $D_p = 12.5\ \text{cm}^2/\text{s}$.

Answer.

$$\tau_d = \frac{10^{-8}}{2 \times 12.5} = 4 \times 10^{-10}\ \text{s}$$

which is indeed very short.

There is also a *circuit arrangement* that speeds up the response. It consists of connecting a capacitance C in parallel with the resistance R and choosing C such that (Fig. 2.16)

$$CR = C_d R_d = \tau \tag{2.27}$$

This speeds up the response because the circuit acts as a simple voltage divider. To demonstrate this, we calculate the frequency response of the circuit for $R_s = 0$ (Fig. 2.16)

$$V_D = V \frac{R_d/(1 + j\omega C_d R_d)}{R/(1 + j\omega CR) + R_d/(1 + j\omega C_d R_d)} = V \frac{R_d}{R + R_d} \tag{2.27a}$$

because $CR = C_d R_d = \tau$. This is independent of frequency, so that $V_D(t)$

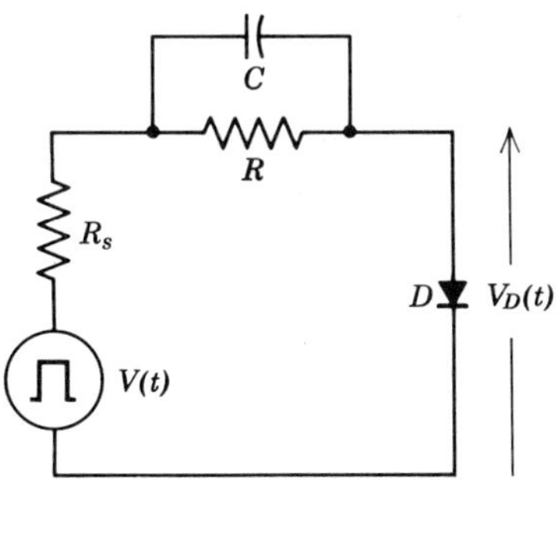

Figure 2.16. Circuit for speeding up the diode response by connecting a properly chosen capacitance C across the resistance R.

follows $V(t)$ instantaneously. Since all reference to the device has disappeared from Eq. (2.27a), this equation holds for arbitrary signals.

The circuit does not work as well in practice, since $V(t)$ is usually developed in series with a finite resistance R_s. Because C, R, and R_s are now in parallel, and the diode impance Z_D is quite small, the new time constant is

$$\tau' = C\frac{RR_s}{R+R_s} = \tau\frac{R_s}{R+R_s} \tag{2.27b}$$

2.4e Effect of the Actual Value of C_d

Since the charge-control approximation assumes $C_d = C_s$, whereas in fact $C_d \cong \frac{1}{2}C_s$, we must now see what influence this has on the response. We shall see that it speeds up the response by a factor of 2.

To demonstrate this, we return to Fig. 2.13. The differential equation for the circuit is, for small signals,

$$C_d\frac{dV_D}{dt} + \frac{V_D}{R_d} = I(t) \tag{2.28}$$

or

$$\frac{1}{2}C_s\frac{dV_D}{dt} + \frac{V_D}{R_d} = I(t) \tag{2.28a}$$

But

$$Q_s = C_s V_D \qquad \text{and} \qquad \frac{V_D}{R_d} = \frac{C_s V_D}{C_s R_d} = \frac{Q_s}{\tau} \tag{2.28b}$$

since $C_s R_d = \tau$. Consequently our equation can be written

$$\frac{1}{2}\frac{dQ_s}{dt} + \frac{Q_s}{\tau} = I(t) \tag{2.29}$$

Now, $Q_s(0^+)$ is given by the initial condition, whereas $Q_s(\infty)$ is obtained by putting $dQ_s/dt = 0$ into Eq. (2.29), bearing into account the appropriate values of $I(t)$ in each case. When $I(t) = I_1$ for $t > 0$ (turn-on transient), we have $Q_s(\infty) = I_1\tau$; when $I(t) = -I_2$ for $t > 0$ (turn-off transient), we have $Q_s(\infty) = -I_2\tau$. The solution to the homogeneous counterpart of Eq. (2.29) is

$$Q_s = A\exp\left(-\frac{2t}{\tau}\right) \tag{2.30}$$

Therefore the full solution is

$$Q_s(t) = Q_s(0^+)\exp\left(-\frac{2t}{\tau}\right) + Q_s(\infty)\left[1 - \exp\left(-\frac{2t}{\tau}\right)\right] \tag{2.31}$$

whereas the charge-control approach in Section 2.4b gives

$$Q_s(t) = Q_s(0^+)\exp\left(-\frac{t}{\tau}\right) + Q_s(\infty)\left[1 - \exp\left(-\frac{t}{\tau}\right)\right] \tag{2.32}$$

This indicates that the transients of Section 2.4b were in error by a factor of 2 in the exponents.

2.5 DIODE SWITCHING EFFECTS IN OR AND AND CIRCUITS

We now show that, while it takes some time to switch a diode, the logic operations of or and and diode circuits are practically instantaneous.

We first look at the or circuit in Fig. 2.17*a*. The input voltage $V(t)$ is a 5-V pulse of sufficiently long duration T. If the pulse is switched on at $t=0$, the output voltage V_0 rises to 5.00 V, since the diode acts as a capacitor and $V_D(0^+)=0$. After that $V_D(t)$ rises to 0.70 V, and the output voltage V_0 drops to 4.30 V. If a 1 corresponds to $V_0 \geqslant 2.0$ V, the logic response is instantaneous.

If the pulse is then turned off at $t=T$, $V_0(t)$ drops instantaneously to -0.70 V, after which it at first rises very slowly, while the stored charge is being removed, and then rises to 0 V when the diode capacitance C_j is discharged. If a 0 output corresponds to $V_0 \leqslant 0.50$ V, the logic response is again instantaneous.

Figure 2.17*b* shows the pulse response for the two cases.

Next we turn to the and circuit in Fig. 2.17*c*. Let $V(t)$ be a 5.0-V pulse starting at $t=0$ and turned off at $t=T$. If the pulse is switched on, then at $t=0^+$, $V_A(0^+)=5.7$ V and $V_0(0^+)\cong 5.0$ V, since $V_{D2}(0^+)\cong 0.70$ V. The output voltage drops gradually to $(5.0-0.70)\times 2000/3000 = 2.87$ V, while the diode D_1 is discharged. If a 1 output corresponds to $V_0(t) \geqslant 2.0$ V, the logic operation is instantaneous.

At $t=T$ the pulse $V(t)$ is turned off. At $t=T^-$, $V_{D1}(T^-)=-1.43$ V, since $V_A(T^-)=2.87+0.70=3.57$ V; hence $V_A(T^+)=-1.43$ V, so $V_0(T^+) = -1.43-0.70 = -2.13$ V. The diode D_1 is now charged, and $V_0(t)$ gradually goes to zero. If a 0 output corresponds to $V_0(t) \leqslant 0.50$ V, the logic operation is again instantaneous.

Figure 2.17*d* shows the pulse response for the two cases.

This does not mean that it is unimportant to have a fast diode response. It is necessary if one wants a high repetition rate for the logic operation.

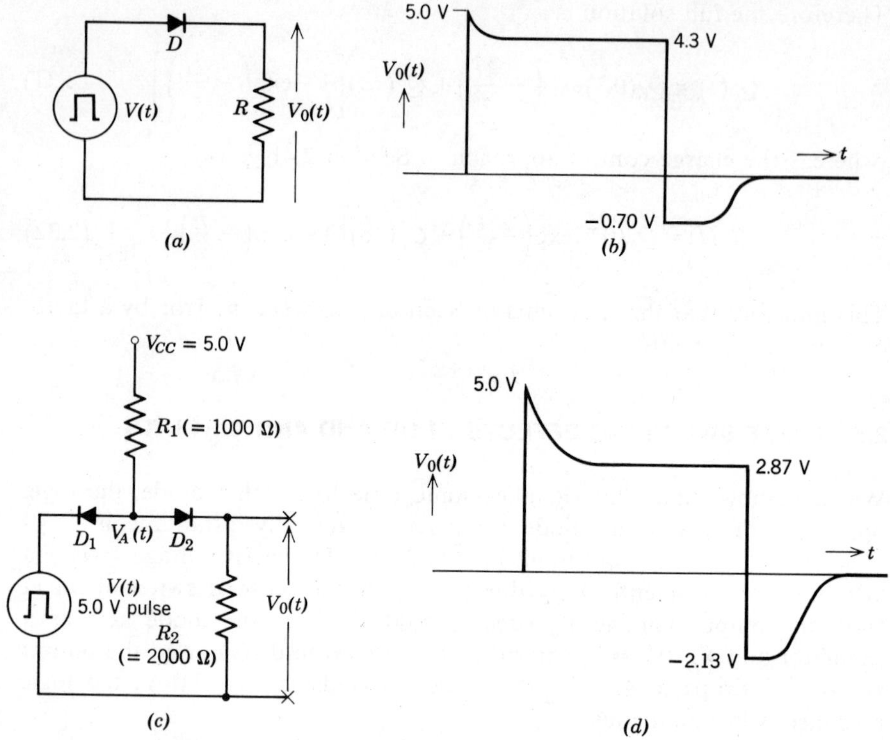

Figure 2.17. (*a*) Part of a diode or circuit. (*b*) Transient response of the circuit, showing the turn-on and turn-off response. (*c*) Part of a diode and circuit. (*d*) Transient response of the circuit showing the turn-on and turn-off response.

PROBLEMS

2.1 In the circuit in Fig. 2.3*a*, $V_{BB} = 1.40$ V, and $R = 5000\ \Omega$. The diode has $V_D = V_{DO} = 0.70$ V at $I_D = I_{DO} = 1.0$ mA. Find the voltage V_D and the current I_D by successive approximation.

2.2 The battery charger in Fig. 2.6 has $V_s(t) = 50 \cos 2\pi ft$ and $R_s = 0.50\ \Omega$, whereas $V_B = 12.0$ V.

(a) Find R so that the maximum current is 30 A.

(b) Find the maximum back bias the diode must be able to withstand.

(c) Find the part of the cycle during which the diode carries current.

(d) Find the average current passing through the battery.

(e) Find how long it takes to charge the fully discharged battery if its capacity is 75 AH.

Hint for (d).

$$I_{av} = \frac{1}{2\pi} \int_{-\pi}^{\pi} I(x)\, dx$$

where $x = \omega t$ is the variable.

2.3 (a) In Problem 2.2 find the average power supplied by $V_s(t)$ during the charging of the battery.
(b) How many kilowatthours are needed to charge the battery?
Hint for (a).

$$P_{av} = \frac{1}{2\pi} \int_{-\pi}^{\pi} P(x)\, dx$$

where $x = \omega t$ is the variable and $P(\omega t)$ is the instantaneous power.

2.4 Design a limiter circuit in Fig. 2.7 that transforms the signal $V_s(t) = 20 \cos \omega t$ V into a trapezoidal waveform with a 10.0 V peak amplitude, bearing in mind that each diode has a turn-on voltage of 0.70 V.

2.5 Plot the voltage transfer characteristic of the circuit in Fig. 2.18 and plot the input current as a function of the input voltage. Locate especially the breakpoints in these characteristics. Neglect the turn-on voltage of the diodes.

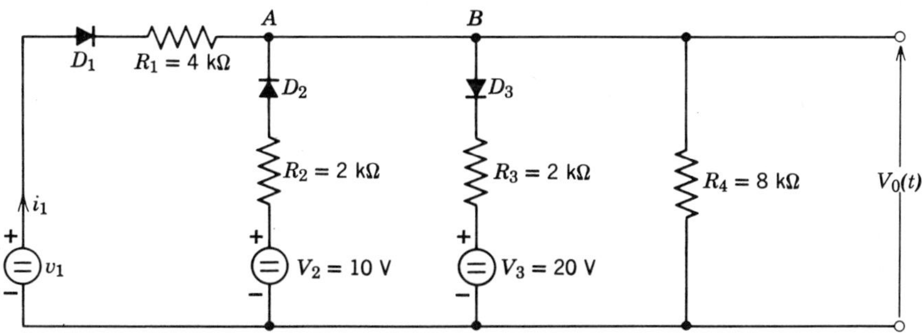

Figure 2.18

2.6 In the or circuit in Fig. 2.10*a*, $V_i(t) = 5.0$ V. Find V_0 if the devices have a turn-on voltage of 0.70 V.

2.7 In the and circuit in Fig. 2.10*b*, $V_B = 5.0$ V and $V_i(t) = 5.0$ V in each channel. Find V_0 if $R_1 = 5$ KΩ and $R_3 = 15$ KΩ. The diodes have a turn-on voltage of 0.70 V.

2.8 In the circuit in Fig. 2.12, $V_s = 10$ V.
(a) Find the output waveform for $V_B = 5$ V, 10 V, and 15 V. Neglect the turn-on voltage of the diode.
(b) Show that the circuit clamps the top of the waveform to V_B if $V_B < 10$ V.

2.9 A silicon p–n diode at T=300°K, having $I_D = 1.0$ mA at $V_D = V_{DO} = 0.70$ V, is used as a small-signal resistor. Find the range of V_D values for which the small-signal resistance varies between 50 and 5000 Ω.

2.10 Find the storage capacitance in a silicon diode carrying a diode current of 1.0 mA if the carrier lifetime is 10^{-7} s.

2.11 In the circuit in Fig. 2.16 the diode has a lifetime of 10^{-7} s. The resistance R in series with the diode is 2000 Ω. Find the capacitance C that must be put in parallel with R to eliminate the turn-on transient.

2.12 In the circuit in Fig. 2.14a the silicon diode has $\tau = 10^{-7}$ s and $I_D = I_{DO} = 1.0$ mA at $V_D = V_{DO} = 0.70$ V.
(a) If a pulse of height $V_1 = 6.0$ V is applied to the diode, find the resistance R for which the maximum current is 1.0 mA.
(b) Find the time interval for which the turn-on voltage transient rises from 0% to 90% of the final value. Neglect the turn-on delay defined in Section 2.4c.

2.13 (a) In the circuit in Fig. 2.14a, $V(t)$ switches from +6.0 V to −6.0 V. Calculate the turn-off transient for $\tau = 10^{-7}$ s if $R = 5300$ Ω.
(b) Repeat the calculation in (a) if $V(t)$ switches from 6.0 V to 0 V.

2.14 In the diode in Problem 2.12, $V(t)$ switches from 0 V to 6.0 V. The diode starts drawing current for $V_D(t) \cong 0.55$ V. The resistance R has the same value as in Problem 2.12a.
(a) Calculate the turn-on delay time of the diode if the input capacitance C_j is 5.0 pF when the diode is not drawing current. The turn-on delay time is defined as the time it takes for $V_D(t)$ to reach a value of 0.55 V.
(b) Repeat (a) if $V(t)$ switches from −6.0 V to +6.0 V.

2.15 Calculate the turn-off transient of the diode in Fig. 2.14a if $I_2 = 0$ and $V_{DO} = 0.70$ V.
Hint. Apply Eq. (2.23a) and substitute $I_2 = 0$ in Eq. (2.20).

2.16 The diode in Fig. 2.14a has been passing 1 mA of current for $t < 0$. At $t = 0$, $V(t)$ is switched to a negative voltage, and a current of −2 mA flows for 10^{-8} s.
(a) Neglecting stored charge decay for $t > 0$, estimate the stored charge at $t = 0$ and the carrier lifetime τ.
(b) Taking the stored charge decay into account, calculate the carrier lifetime τ by applying Eq. (2.22).

(c) Knowing the lifetime τ from (b) above, calculate the stored charge more accurately.

2.17 In the circuit in Fig. 2.19 find and sketch:

(a) The turn-on transient $V_1 = 5$ V and $V_2 = 0$ V
(b) The turn-off transient $V_1 = 5$ V and $V_2 = 0$ V
if $V_{DO} = 0.70$ V at $I_{DO} = 1.0$ mA, and $\tau = 10^{-8}$ s. In both cases neglect the space-charge capacitance C_j of the diode D.

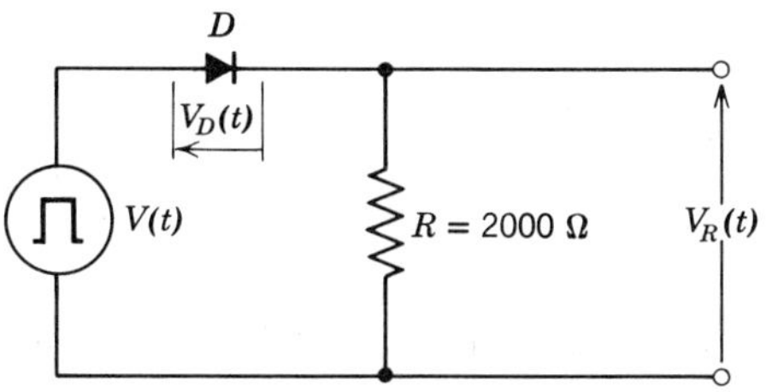

Figure 2.19

2.18 Draw a circuit using ideal diodes, resistors, voltage, and current sources that will give the I–V diagrams shown in Fig. 2.20. Note that the answer is not unique, and that many solutions are possible.

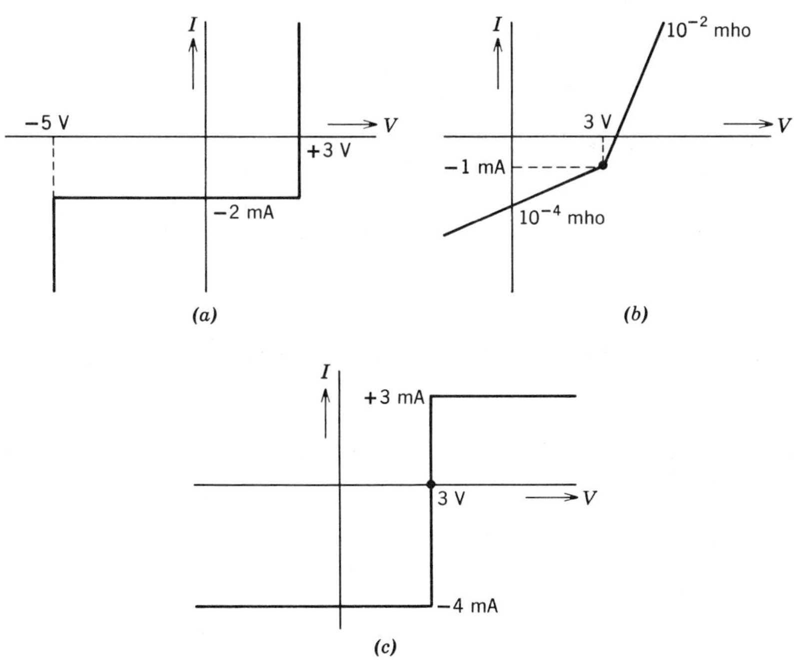

Figure 2.20

CHAPTER

3 ___ Nonlinear Transistor Models and Circuits

In this chapter we discuss the pulse response of transistor circuits. Section 3.1 describes the application of the Ebers-Moll equations to large-signal transistor operations. Section 3.2 considers two large-signal transistor circuits, and Section 3.3 deals with the transient response when these circuits operate into a capacitive load.

Section 3.4 discusses the intrinsic response of transistor inverters using the charge-control approach. The influence of the Miller effect on the transient response is investigated. It is shown that an overdriving of the base results in a turn-off delay. The response during the inactive part of the transistor transient is investigated; the effect of the junction capacitances results in a turn-on delay. Section 3.5 considers the influence of the Early effect on the transient response.

3.1 THE EBERS–MOLL EQUATIONS AND THEIR APPLICATIONS

The large-signal behavior of transistors is best described by the Ebers–Moll equations. We first derive these equations and then apply them to calculation of the collector voltage under saturated conditions.

3.1a The Ebers–Moll Equations

For a normally biased $p-n-p$ transistor, neglecting small saturation currents, the current equations are

$$I_E = I_{ES} \exp\left(\frac{eV_{EB}}{kT}\right) \qquad I_C = -\alpha_F I_{ES} \exp\left(\frac{eV_{EB}}{kT}\right) \tag{3.1}$$

where V_{EB} is the dc bias voltage applied between the emitter and the base and I_{ES} is a kind of saturation current. Furthermore, α_F is called the *forward current amplification factor*. If all current is carried by holes, it represents the part of the hole current injected by the emitter that is collected by the collector; normally, however, some electrons also flow from the base to the emitter. Here the directions of positive current flow are *into* the emitter, *into* the collector, and *into* the base.

For modern transistors I_{ES} is of the order of 10^{-15} A. Hence, for an emitter current $I_E = 1$ mA,

$$V_{EB} = \frac{kT}{e} \ln\left(\frac{I_E}{I_{ES}}\right) = 25.8 \times 10^{-3} \ln 10^{12} = 0.71 \text{ V}$$

The emitter-base voltage for a 1-mA emitter current is called the *turn-on voltage* of the transistor. Unless otherwise stated, we assume a turn-on voltage of 0.70 V.

In some pulse applications the emitter-base junction and the collector-base junction are *both* forward-biased. We now investigate the current flow for this situation in two steps, again for a $p-n-p$ transistor.

In the first step we reverse the bias voltages; that is, the collector-base junction is now forward-biased and the emitter-base junction is back-biased. This is called *reverse* operation. In analogy with Eqs. (3.1) we have

$$I_C = I_{CS} \exp\left(\frac{eV_{CB}}{kT}\right) \qquad I_E = -\alpha_R I_{CS} \exp\left(\frac{eV_{CB}}{kT}\right) \tag{3.2}$$

Here V_{CB} is the (forward) collector-base voltage, I_{CS} is a kind of saturation current, and α_R is the *reverse current amplification factor*. If all current is carried by holes, α_R represents the part of the hole current injected by the collector that is collected by the emitter; usually, however, an appreciable electron current flows from the base to the collector in this bias configuration.

In the second step both junctions are forward-biased, that is, V_{EB} and V_{CB} are both positive. The current is found by combining Eqs. (3.1) and

(3.2). Hence

$$I_E = I_{ES}\exp\left(\frac{eV_{EB}}{kT}\right) - \alpha_R I_{CS}\exp\left(\frac{eV_{CB}}{kT}\right) \tag{3.3}$$

$$I_C = -\alpha_F I_{ES}\exp\left(\frac{eV_{EB}}{kT}\right) + I_{CS}\exp\left(\frac{eV_{CB}}{kT}\right) \tag{3.4}$$

Since $I_B + I_E + I_C = 0$, the base current I_B is

$$I_B = -(1-\alpha_F)I_{ES}\exp\left(\frac{eV_{EB}}{kT}\right) - (1-\alpha_R)I_{CS}\exp\left(\frac{eV_{CB}}{kT}\right) \tag{3.5}$$

Equations (3.3) through (3.5) are called the *Ebers–Moll equations* for the *p–n–p* transistor.

Similar equations hold for the silicon $n-p-n$ transistor, except that the directions of all currents and the polarities of all voltages are reversed. We thus have, if $V_{BE} = -V_{EB}$ and $V_{BC} = -V_{CB}$ are the base-emitter and base-collector voltages, respectively,

$$I_E = -I_{ES}\exp\left(\frac{eV_{BE}}{kT}\right) + \alpha_R I_{CS}\exp\left(\frac{eV_{BC}}{kT}\right) \tag{3.6}$$

$$I_C = \alpha_F I_{ES}\exp\left(\frac{eV_{BE}}{kT}\right) - I_{CS}\exp\left(\frac{eV_{BC}}{kT}\right) \tag{3.7}$$

$$I_B = (1-\alpha_F)I_{ES}\exp\left(\frac{eV_{BE}}{kT}\right) + (1-\alpha_R)I_{CS}\exp\left(\frac{eV_{BC}}{kT}\right) \tag{3.8}$$

It can be shown on general grounds that*

$$\alpha_F I_{ES} = \alpha_R I_{CS} \tag{3.9}$$

This is called the *Onsager relation* for transistors.

Something further should be said about the parameters α_F, α_R, I_{ES}, and I_{CS}. Most $p-n-p$ transistors are of the $p^+-n-p^--p^+$ variety. That is, they are made on a p^+-wafer with an epitaxial, weakly p-type layer (p^--layer) grown on it; afterward the n-type base region and the p^+-type emitter region are diffused in. Because the emitter-base junction is of the

*A. van der Ziel, *Solid State Physical Electronics*, 3rd Ed., Prentice Hall, Englewood Cliffs, N.J., 1976, Chapter 16.

p^+-n type, practically all emitter current is hole current. Since the base region is very narrow, practically all injected holes are collected by the collector, so that α_F is very close to unity. Because the base-collector junction is of the $n-p^-$ variety, a major part of the collector current in Eq. (3.2) may be *electron* current. Since only the injected *holes* are collected by the emitter, α_R can be quite small; values as low as 0.1 are possible. This is enhanced by the fact that the collector area is much larger than the base area (Fig. 3.1), so that many injected holes never find the emitter.

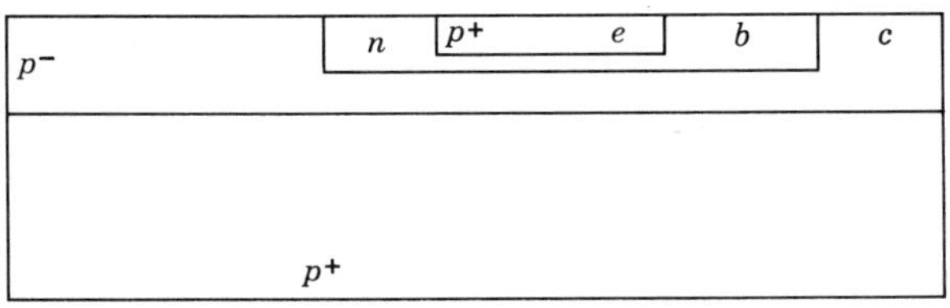

Figure 3.1. Cross section of a p–n–p transistor on a p^+-substrate.

In pulse applications one usually operates the transistor in either one of two regimes:

1. The transistor is off. This does not require application of the Ebers–Moll equations, one simply has $I_C = 0$.

2. The collector is *saturated*, that is, the collector current is determined by the external circuit and V_{CE} is close to 0 V. The Ebers–Moll equations can be used to evaluate the value of V_{CE}.

It is convenient to show polarities of the voltages in a figure. We do so for a p–n–p transistor in Fig. 3.2*a* and for an n–p–n transistor in Fig. 3.2*b*.

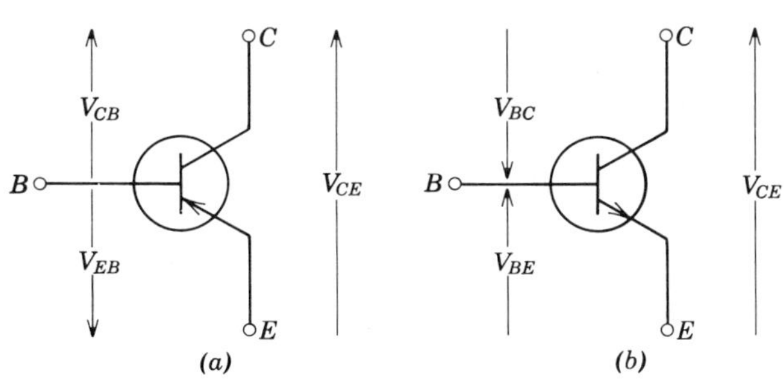

Figure 3.2. (*a*) Symbol for a p–n–p transistor. (*b*) Symbol for an n–p–n transistor.

3.1b Calculation of V_{CE} under Saturated Conditions

As a preliminary to calculating V_{CE} under saturated conditions we first give two examples:

Example 1. In a *p–n–p* transistor with a forward-biased emitter-base junction the collector is left floating. Find the potential difference V_{CE} between the floating collector and the emitter, in particular for $\alpha_R = 0.10$.

Answer. As seen from Fig. 3.2 we have

$$V_{CB} = V_{CE} + V_{EB} \tag{3.10}$$

Substituting into Eq. (3.4) and putting $I_C = 0$ yields

$$0 = -\alpha_F I_{ES} \exp\left(\frac{eV_{EB}}{kT}\right) + I_{CS} \exp\left(\frac{eV_{EB}}{kT}\right) \exp\left(\frac{eV_{CE}}{kT}\right)$$

so that

$$\exp\left(\frac{eV_{CE}}{kT}\right) = \frac{\alpha_F I_{ES}}{I_{CS}} = \alpha_R \qquad V_{CE} = \frac{kT}{e} \ln \alpha_R$$

because of Eq. (3.9). Since $\alpha_R < 1$, V_{CE} is slightly negative. For $\alpha_R = 0.10, V_{CE} \cong -0.060$ V.

Example 2. Repeat the calculation in Example 1 for an *n–p–n* transistor with $\alpha_R = 0.10$.

Answer. V_{CE} has a value opposite that in the previous case, that is,

$$V_{CE} = -\frac{kT}{e} \ln \alpha_R$$

so that $V_{CE} \cong 0.060$ V if $\alpha_R = 0.10$.

We now calculate the value of V_{CE} for a saturated *n–p–n* transistor. Starting from the Ebers–Moll equations (3.6) through (3.8), we bear in mind that the collector and emitter junctions are both forward-biased. We therefore introduce

$$I_F = I_{ES} \exp\left(\frac{eV_{BE}}{kT}\right) \qquad I_R = I_{CS} \exp\left(\frac{eV_{BC}}{kT}\right) \tag{3.11}$$

and rewrite the Ebers–Moll equations as

$$I_C = \alpha_F I_F - I_R \qquad I_B = (1-\alpha_F) I_F + (1-\alpha_R) I_R \tag{3.12}$$

Solving for I_F and I_R yields

$$I_F = \frac{(1-\alpha_R) I_C + I_B}{1-\alpha_F\alpha_R} \qquad I_R = \frac{-(1-\alpha_F) I_C + \alpha_F I_B}{1-\alpha_F\alpha_R} \tag{3.12a}$$

Now, according to Eq. (3.11),

$$V_{BE} = \frac{kT}{e}\ln\left(\frac{I_F}{I_{ES}}\right) \qquad V_{BC} = \frac{kT}{e}\ln\left(\frac{I_R}{I_{CS}}\right) \tag{3.13}$$

But, as is seen from Fig. 3.2,

$$V_{BE} = V_{BC} + V_{CE} \qquad \text{or} \qquad V_{BC} = V_{BE} - V_{CE} \tag{3.14}$$

Since $\alpha_R I_{CS} = \alpha_F I_{ES}$, we have

$$V_{BC} = \frac{kT}{e}\ln\left(\frac{\alpha_R I_R}{\alpha_R I_{CS}}\right) = \frac{kT}{e}\ln\left(\frac{\alpha_R I_R}{\alpha_F I_{ES}}\right)$$

$$= \frac{kT}{e}\ln\left(\frac{I_F}{I_{ES}}\right) + \frac{kT}{e}\ln\left(\frac{\alpha_R I_R}{\alpha_F I_F}\right) = V_{BE} + \frac{kT}{e}\ln\left(\frac{\alpha_R I_R}{\alpha_F I_F}\right) = V_{BE} - V_{CE}$$

according to Eq. (3.14). Hence

$$V_{CE} = -\frac{kT}{e}\ln\left(\frac{\alpha_R I_R}{\alpha_F I_F}\right) = \frac{kT}{e}\ln\left(\frac{\alpha_F I_F}{\alpha_R I_R}\right) \tag{3.15}$$

It is often convenient to express V_{CE} directly in terms of I_C and I_B with the help of Eq. (3.12a). This yields

$$V_{CE} = \frac{kT}{e}\ln\frac{\alpha_F[(1-\alpha_R) I_C + I_B]}{\alpha_R[-(1-\alpha_F) I_C + \alpha_F I_B]}$$

$$= \frac{kT}{e}\ln\left[\frac{(1-\alpha_R) I_C + I_B}{\alpha_R(-I_C/\beta_F + I_B)}\right] \tag{3.16}$$

where $\beta_F = \alpha_F/(1-\alpha_F)$ is the current amplification factor of an unsaturated forward-biased transistor in common emitter configuration.

According to Eqs. (3.6) through (3.8), we have in that case $\alpha_F = -I_C/E_E$ and $I_B = -(I_E + I_C)$, so

$$\beta_F = \frac{\alpha_F}{1-\alpha_F} = \frac{-I_C/I_E}{1+I_C/I_E} = \frac{I_C}{-(I_E+I_C)} = \frac{I_C}{I_B} \tag{3.16a}$$

Hence, for a forward-biased unsaturated transistor Eq. (3.16) has no meaning. For a saturated transistor $I_B > I_C/\beta_F$, and Eq. (3.16) has meaning.

Equation (3.16) indicates that for $V_{CE} \cong 0.10$–0.20 V there must be an overdrive, that is, I_B must be somewhat larger than I_C/β_F. In the particular case that $I_B \gg I_C/\beta_F$ and α_R is relatively small,

$$V_{CE} \cong \frac{kT}{e} \ln\left(\frac{I_C + I_B}{\alpha_R I_B}\right) \tag{3.16b}$$

Example 3. Find V_{CE} if $\alpha_R = 0.10$ and $I_C = 4I_B$ at $T = 300°$ K.

Answer. $V_{CE} \cong 25.8 \times 10^{-3} \ln 50 = 0.10$ V.

Usually V_{CE} lies between 0.10 and 0.20 V; its value is denoted by $V_{CE,\text{sat}}$.

3.1c Connection of a Transistor as a Diode

There are several ways to connect a transistor as a diode

1. Connect the base and the collector.
2. Connect the base and the emitter.
3. Connect the emitter and the collector.
4. Float the emitter.
5. Float the collector.

In the last four configurations the collector is forward-biased when the device is drawing current. A large minority carrier concentration is then stored in the base region, and this slows up the response. In the first configuration all the minority carrier charge is stored between the emitter and the collector, and it takes only a very short time to remove it [see Eq. (2.26)]. Therefore connection 1 gives the fastest diode response.

3.2 LARGE-SIGNAL TRANSISTOR CIRCUITS

Here we discuss two large-signal transistor circuits. The first is a circuit with a collector load resistance (Section 3.2a), and the second is an emitter follower circuit (Section 3.2c). Both are useful as logic circuits; the first acts as an inverter circuit, and the second transmits the information at the input. The pulse response of these circuits when they operate into a capacitive load is discussed in Section 3.3.

3.2a A Large-Signal Transistor Circuit as a Linear Circuit

Figure 3.3*a* shows a large-signal transistor circuit. We assume that V_{CC} is relatively large, and that V_{BB} is gradually increased from 0 V. As long as

$$V_{BB} < V_{BE0} \cong 0.70 \text{ V} \tag{3.17}$$

the device hardly draws any current, and $V_{CE} \cong V_{CC}$.

For $V_{BB} > 0.70$ V a base current

$$I_B = \frac{V_{BB} - V_{BE0}}{R_b} \tag{3.18}$$

flows, where $V_{BE0} \cong 0.70$ V. The resulting collector current is

$$I_C = \beta_F I_B = \beta_F \frac{V_{BB} - V_{BE0}}{R_b} \tag{3.19}$$

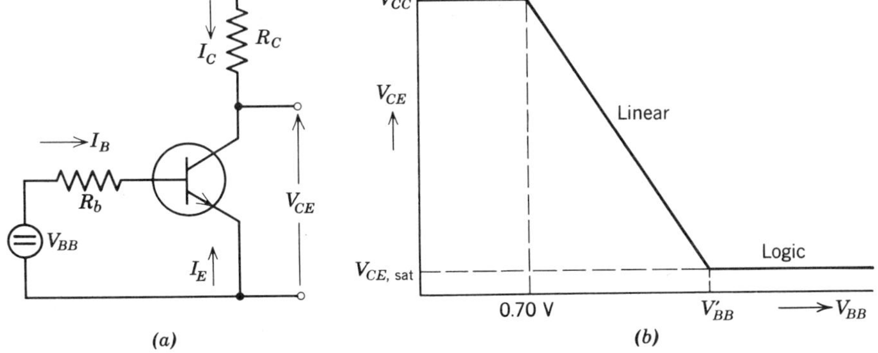

Figure 3.3. (*a*) Large-signal transistor circuit. (*b*) Transfer characteristic of the circuit, showing the linear and the logic regimes.

where β_F is the common emitter current amplification factor for forward bias, so that

$$V_{CE} = V_{CC} - I_C R_C = V_{CC} - \frac{\beta_F R_c}{R_b}(V_{BB} - V_{BE0}) \tag{3.20}$$

The output voltage V_{CE} thus drops linearly with increasing V_{BB}, unless V_{CE} becomes so small that the collector-base junction is forward-biased. The *linear* mode of operation of the circuit is thus the mode in which V_{CE} varies linearly with V_{BB}.

What happens if V_{BB} is further increased? The voltage V_{CE} then remains practically independent of the current, and the transistor is said to be *saturated*. The value of V_{CE} under these conditions is denoted by $V_{CE,\mathrm{sat}}$ and is of the order of 0.10–0.20 V. The current equations are then

$$I_B = \frac{V_{BB} - V_{BE0}}{R_b} \qquad I_C = \frac{V_{CC} - V_{CE,\mathrm{sat}}}{R_c} = I_{Cs} \tag{3.21}$$

and $I_C < \beta_F I_B$. The relationship $I_C = \beta_F I_B$ therefore cannot be used in saturation. What remains true, however, is that

$$-I_E = I_B + I_C \tag{3.22}$$

if the polarities are as shown in Fig. 3.3*a*.

What are the limits of the linear mode of operation? One limit is $V_{BB} = V_{BE0}$; the transistor is then turned off. The other limit is found when the transistor is just saturated, that is, when V_{CE} in Eq. (3.20) has just reached the value $V_{CE,\mathrm{sat}}$. At that point it is still true that $I_C = \beta_F I_B$, so that Eq. (3.20) remains valid. Therefore, inverting this equation,

$$V_{BB} = V'_{BB} = V_{BE0} + (V_{CC} - V_{CE,sat})\frac{R_b}{\beta_F R_c} \tag{3.23}$$

The linear range of operation thus extends over

$$V_{BE0} \leqslant V_{BB} \leqslant V_{BE0} + \frac{R_b}{\beta_F R_c}(V_{CC} - V_{CE,\mathrm{sat}}) \tag{3.23a}$$

The corresponding range of V_{CE} is

$$V_{CC} \geqslant V_{CE} \geqslant V_{CE,\mathrm{sat}} \tag{3.23b}$$

The width of the linear range of operation is given by Eq. (3.23a). This also corresponds to the difference between the off and the saturated conditions.

Figure 3.3*b* shows the transfer characteristic, indicating the two logic ranges $V_{BB} < V_{BE0}$ and $V_{BB} > V'_{BB}$ and the linear range in between.

Example 1. Find V'_{BB} if $R_b = 6R_c$, $\beta_F = 30$, and $V_{CC} = 5.0$ V, $V_{CE,\text{sat}} = 0.10$ V, and $V_{BE0} = 0.70$ V.

Answer. $V'_{BB} = 0.70 + \frac{1}{5} \times 4.9 = 1.68$ V.

3.2b The Circuit as a Logic Circuit

Let $V_{BB} < V_{BE0}$ correspond to the logic 0 condition at the input; then the output voltage $V_{CE} \cong V_{CC}$ will be high, corresponding to the logical 1 condition at the output. Let

$$V_{BB} > V_{BE0} + \frac{(V_{CC} - V_{CE,\text{sat}})R_b}{\beta_F R_c} \tag{3.23c}$$

correspond to a logical 1 condition at the input; then the output voltage $V_{CE} = V_{CE,\text{sat}}$, corresponding to a logical 0 at the output.

The transistor is thus a logical *inverter*, converting a logical 1 into a logical 0, and vice versa. The only requirements are that $V_{CE,\text{sat}} < V_{BE0}$, and that V_{BB} in Eq. (3.23c) be equal to or smaller than V_{CC}. The first condition is always satisfied, since $V_{BE0} \cong 0.70$ V and $V_{CE,\text{sat}} \cong 0.10$–$0.20$ V. The latter condition requires $R_b / \beta_F R_c < 1$, which means that the transistor is saturated; this is also not difficult to satisfy.

Especially in some types of integrated transistors, the collector region has an appreciable series resistance r_c. In this case we have, at saturation,

$$V_{CE} = V_{CE,\text{sat}} + I_{Cs} r_c \tag{3.24}$$

This is especially important in inverter circuits in which an appreciable current I_0 flows into the collector from subsequent stages. In this case the ohmic voltage drop in r_c can give rise to a large increase in the collector voltage V_{CE}. The collector remains saturated as long as

$$I_C = \frac{V_{CC} - V_{CE}}{R_c} + I_0 < \beta_F I_B \tag{3.24a}$$

but the collector voltage is

$$V_{CE} = V_{CE,\text{sat}} + I_C r_c \tag{3.24b}$$

To ensure that the output remains a 0, this voltage must be smaller than

$V_{BE0}=0.70$ V. This seriously limits the amount of current I_0 that can be fed into the collector. This is an important consideration in transistor logic output circuits carrying a large collector current (see Chapter 7).

3.2c The Emitter Follower as a Large-Signal Circuit

Figure 3.4 shows an emitter follower circuit; here a signal V_{BB} is applied to the base via a resistor R_b, and the output is taken from a resistor R_e in the emitter lead. We shall see that the output voltage follows the input voltage linearly, hence the name of the circuit.

If $V_{BB}<V_{BE0}\cong 0.70$ V, the transistor is off and the output voltage V_E is zero. If now V_{BB} is increased above $V_{BE0}\cong 0.70$ V, a current I_B flows, so

$$V_{BB}-V_E-V_{BE0}=R_bI_B \tag{3.25}$$

But

$$V_E=I_ER_e=(\beta_F+1)R_eI_B \tag{3.26}$$

as long as the transistor is not saturated. Substituting into Eq. (3.25) yields

$$I_B[R_b+(\beta_F+1)R_e]=V_{BB}-V_{BE0} \tag{3.27}$$

or, since $I_B=I_E/(\beta_F+1)$,

$$I_E\left[\frac{R_b}{\beta_F+1}+R_e\right]=V_{BB}-V_{BE0} \tag{3.28}$$

Equations (3.27) and (3.28) can be represented by the equivalent circuits in Fig. 3.4*b* and *c*. We thus see from Fig. 3.4*b* that the input circuit "sees" an input resistance $R_e(1+\beta_F)$ in series with the emf V_{BE0} (turn-on voltage), and from Fig. 3.4*c* that the output load resistance R_e "sees" an emf $V_{BB}-V_{BE0}$ in series with an "internal resistance" $R_b/(1+\beta_F)$. This indicates that the circuit transforms the emitter resistance R_e "up" and the base resistance R_b "down," that is, it transforms from a higher to a lower impedance level without loss of signal. We also see that

$$V_E=(V_{BB}-V_{BE0})\frac{R_e}{R_e+R_b/(1+\beta_F)} \tag{3.29}$$

so the output voltage V_E follows the emf V_{BB} linearly as long as the

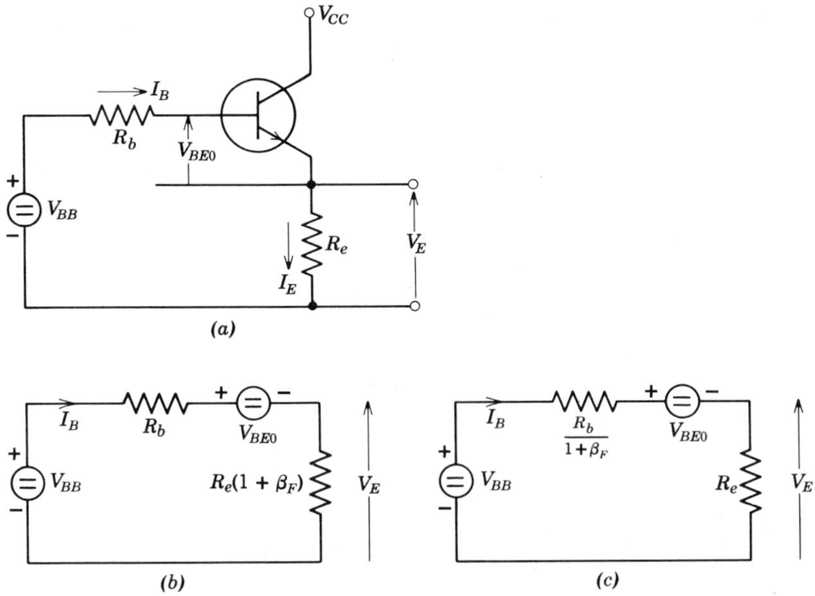

Figure 3.4. (*a*) Large-signal emitter follower circuit. (*b*) Equivalent circuit as seen from the base side. (*c*) Equivalent circuit as seen from the emitter side.

transistor is on ($V_{BB} > V_{BE0}$) and is not saturated ($V_E < V_{CC} - V_{CE,\text{sat}}$). The upper limit of the output voltage is set by collector saturation,

$$V_E = V_{CC} - V_{CE,\text{sat}} \tag{3.30}$$

for which the corresponding value of V_{BB} is

$$V_{BB} = V_{BE0} + (V_{CC} - V_{CE,\text{sat}})\frac{R_e + R_b/(1+\beta_F)}{R_e} \tag{3.31}$$

The range of linear operation for V_{BB} is thus

$$V_{BE0} \leqslant V_{BB} \leqslant V_{BE0} + (V_{CC} - V_{CE,\text{sat}})\frac{R_e + R_b/(1+\beta_F)}{R_e} \tag{3.32}$$

corresponding to a range of output voltage

$$0 \leqslant V_E \leqslant V_{CC} - V_{CE,\text{sat}} \tag{3.33}$$

The circuit can also be used as a logic circuit. A 0 input ($V_{BB} < V_{BE0}$) corresponds to a 0 output, and a 1 input ($V_{BB} \cong V_{CC}$) corresponds to a 1 output. *The circuit therefore transmits the logic information without altering it.*

3.3 TRANSIENTS IN BASIC TRANSISTOR SWITCHING CIRCUITS

Here we evaluate the transient response of the transistor inverter circuit and the emitter follower circuit. We assume that these circuits feed into other circuits, and that their load impedance can be represented by a capacitance C_0. The intrinsic response of transistors is discussed in Section 3.4.

3.3a The Transistor Inverter Circuit Operating into a Capacitive Load

Figure 3.5 shows the inverter circuit with a capacitance C_0 (load capacitance) connected between the collector and the ground. If V_c is the voltage developed across C_0, the turn-on transient satisfies the differential equation

$$C_0 \frac{dV_c}{dt} + \beta_F I_B = \frac{V_{CC} - V_c}{R_c} \tag{3.34}$$

since $I_C = \beta_F I_B$ as long as the collector current is unsaturated. The equation holds as long as

$$V_c > V_{CE,\text{sat}}$$

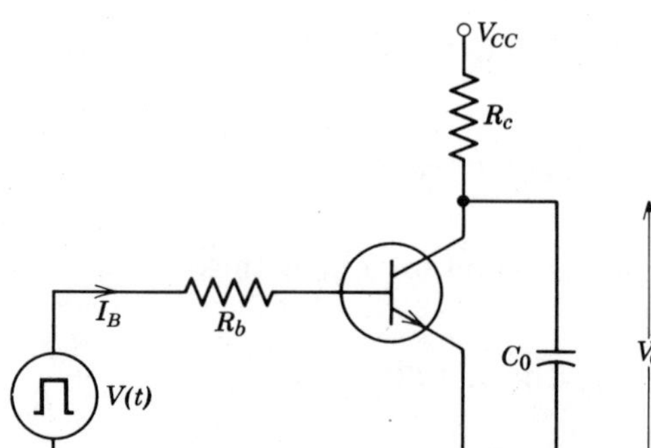

Figure 3.5. Large-signal transistor circuit with a load capacitor C_0 connected between the collector and the ground.

Dividing by the capacitance C_0 and introducing the time constant $\tau = C_0 R_c$, we can write Eq. (3.34) as

$$\frac{dV_c}{dt} + \frac{V_c}{\tau} = \frac{V_{CC} - \beta_F I_B R_c}{\tau} \tag{3.34a}$$

with the initial condition $V_c(0^+) = V_{CC}$, since the transistor is off for $t < 0$. But

$$V_c(\infty) = V_{CC} - \beta_F I_B R_c \tag{3.34b}$$

hence the solution to Eq. (3.34a) can be written

$$V_c(t) = V_{CC} \exp\left(-\frac{t}{\tau}\right) + (V_{CC} - \beta_F I_B R_c)\left[1 - \exp\left(-\frac{t}{\tau}\right)\right] \tag{3.35}$$

The turn-on transient ends when $V_c(t) = V_{CE,\text{sat}}$, or

$$V_{CC} - \beta_F I_B R_c + \beta_F I_B R_c \exp\left(-\frac{t}{\tau}\right) = V_{CE,\text{sat}}$$

or

$$\exp\left(\frac{t}{\tau}\right) = \frac{\beta_F I_B R_c}{\beta_F I_B R_c - V_{CC} + V_{CE,\text{sat}}}$$

or

$$t = t_1 = \tau \ln\left[\frac{\beta_F I_B R_c}{\beta_F I_B R_c - V_{CC} + V_{CE,\text{sat}}}\right] \tag{3.36}$$

Since $\beta_F I_B R_c \gg V_{CC} - V_{CE,\text{sat}}$, the turn-on transient is quite short. We can approximate $-\ln(1-x)$ by x for small x, hence we can write

$$t_1 = -\tau \ln\left[1 - \frac{V_{CC} - V_{CE,\text{sat}}}{\beta_F I_B R_c}\right] \cong C_0 \frac{(V_{CC} - V_{CE,\text{sat}})}{\beta_F I_B} \tag{3.36a}$$

This has the following simple interpretation. Initially, the capacitor has a charge $Q_0 = C_0 V_{CC}$, the discharge current $I_C = \beta_F I_B$, and the final charge in $C_0 V_{CE,\text{sat}}$; hence the discharge time is $C_0(V_{CC} - V_{CE,\text{sat}})/I_C$, in agreement with Eq. (3.36a). In some cases, however, the full Eq. (3.36) should be used.

The turn-off transient lasts much longer. Since the transistor is off for $t>0$,

$$C_0\frac{dV_c}{dt}=\frac{V_{CC}-V_c}{R_c} \qquad \text{or} \qquad \frac{dV_c}{dt}+\frac{V_c}{\tau}=\frac{V_{CC}}{\tau} \tag{3.37}$$

with $V_c(0^+)\cong 0$*. We read from Eq. (3.37) that $V_c(\infty)=V_{CC}$, so the turn-off transient is

$$V_c(t)=V_{CC}\left[1-\exp\left(-\frac{t}{\tau}\right)\right] \tag{3.38}$$

and 90% of the transient has been completed if

$$t=t_2=2.3\tau \tag{3.38a}$$

Example 1. If $C_0=50$ pF, $R_c=2000\ \Omega$, $V_{CC}=10$ V, $V_{CE,\text{sat}}=0.0$ V, $I_B=1$ mA, and $\beta_F=100$, determine t_1 and t_2.

Answer.

$$\tau=C_0R_c=50\times10^{-12}\times2000=10^{-7}\text{ s}$$

$$\beta_F I_B R_c=100\times10^{-3}\times2000=200\text{ V}$$

$$t_1=\tau\ln\left(\frac{\beta_F I_B R_c}{\beta_F I_B R_c-V_{CC}}\right)=10^{-7}\ln\left(\frac{200}{190}\right)=0.51\times10^{-8}\text{ s}$$

so t_1 is quite short. If Eq. (3.36a) had been used instead, we would have obtained

$$t_1=\frac{50\times10^{-12}\times10}{100\times10^{-3}}=0.50\times10^{-8}\text{ s}$$

which is quite close. Furthermore, $t_2=2.3\times10^{-7}$ s.

3.3b The Emitter Follower Circuit with a Capacitive Load

Figure 3.6*a* shows the circuit. The load is again represented by a capacitance C_0. For the turn-on transient, $V(t)$ switches from $0V$ to $+V_1$ at $t=0$; for the turn-off transient, $V(t)$ switches from $+V_1$ to $0V$ at $t=0$.

*We take here $V_{CE,\text{sat}}\cong 0$.

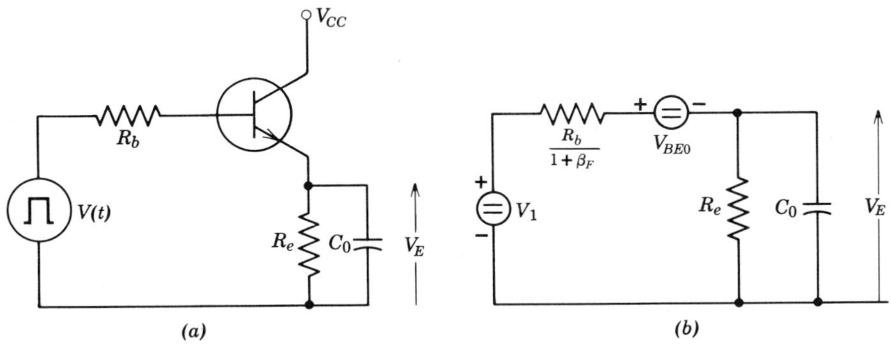

Figure 3.6. (*a*) Large-signal emitter follower circuit with a load capacitor C_0 connected across the output. (*b*) Equivalent circuit as seen from the emitter side.

Since V_1 usually comes from a previous stage fed from the same supply voltage V_{CC}, we have $V_1 \leqslant V_{CC}$. Since V_E has a maximum value $V_1 - V_{BE0}$, where $V_{BE0} \cong 0.70$ V, we see that $V_{CE} = V_{CC} - V_E \geqslant 0.70$ V, so the collector of the transistor is never saturated.

The equivalent circuit for the turn-on transient is given by Fig. 3.6*b*, with $V_E(0^+)=0$. As seen by inspection,* $V_E(\infty) \cong V_1 - V_{BE0}$ if $R_b/(1+\beta_F) \ll R_E$. Moreover, the time constant τ_1 of the circuit under these conditions is

$$\tau_1 = \frac{C_0 R_b}{1+\beta_F} \tag{3.39}$$

Hence the turn-on transient is

$$V_E(t) \cong (V_1 - V_{BE0})\left[1 - \exp\left(-\frac{t}{\tau_1}\right)\right] \tag{3.40}$$

so the transient is 90% complete if

$$t = t_1 = 2.3\tau_1 \tag{3.41}$$

For the turn-off transient the transistor is off for $t>0$. Since the voltage on the capacitor at $t=0^+$ is $V_c(0^+) = V_1 - V_{BE0}$, the voltage decays freely with a time constant $\tau_2 = C_0 R_c$, so

$$V(t) = (V_1 - V_{BE0})\exp\left(-\frac{t}{\tau_2}\right) \tag{3.42}$$

*Actually, there is a small voltage drop $I_B R_b$ in the resistance R_b, but its effect is negligible if $R_b/(1+\beta_F) \ll R_E$.

The transient is 90% complete if

$$t = t_2 = 2.3\tau_2 \tag{3.43}$$

Since $\tau_1 \ll \tau_2$, the turn-on transient is much faster than the turn-off transient.

Example 2. If $R_b = 5000\ \Omega$, $R_e = 1000\ \Omega$, $C_0 = 50$ pF, and $\beta_F = 50$, find t_1 and t_2.

Answer. $\tau_1 = 50 \times 10^{-12} \times (5000/51) \cong 5 \times 10^{-9}$ s; $t_1 = 2.3\tau_1 \cong 1.1 \times 10^{-8}$ s. $\tau_2 = 50 \times 10^{-12} \times 1000 = 5 \times 10^{-8}$ s; $t_2 = 2.3\tau_2 \cong 11 \times 10^{-8}$ s.

Since the transistor inverter and the emitter follower both have a very short turn-on time, circuits have been designed that combine the favorable properties of both (totem-pole circuits; see Chapter 7).

3.4 INTRINSIC RESPONSE OF TRANSISTOR INVERTERS

Thus far we have assumed that the transistor itself responds instantaneously. Actually, the response time of the transistor is finite, and in most cases this must be taken into account. We do so for an *n–p–n* transistor.

Figure 3.7 shows the equivalent circuit for the transistor, in which the series resistance r_x of the base has been omitted. The response time is finite because the equivalent circuit contains two capacitors: the capacitance C_π in parallel with the resistance r_π and the capacitance C_μ between the base and the collector. The first gives rise to an input time constant $\tau_\pi = C_\pi r_\pi$, and the second gives rise to a Miller-effect capacitance which slows up the response even more.

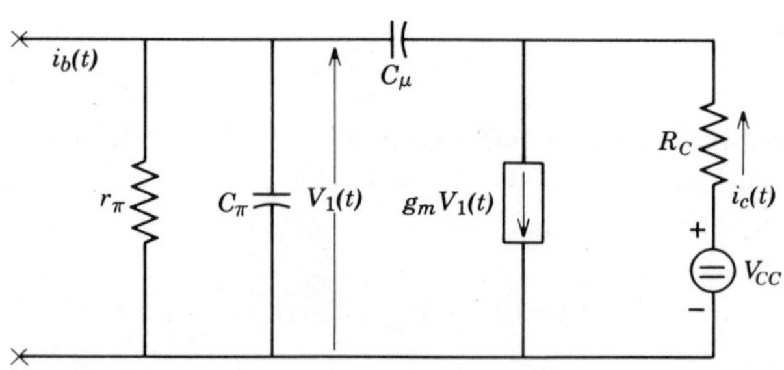

Figure 3.7. Equivalent circuit of the transistor, showing the feedback capacitance C_μ and the input admittance consisting of a resistance r_π and a capacitance C_π in parallel.

The input resistance r_π comes about as follows. Since I_B and I_C depend on V_{BE} in the manner described by Eqs. (3.7) and (3.8), we can introduce the input conductance g_{be} and the transconductance g_m with the equations

$$g_{be}=\frac{\partial I_B}{\partial V_{BE}}=\frac{eI_B}{kT} \qquad g_m=\frac{\partial I_C}{\partial V_{BE}}=\frac{eI_C}{kT} \tag{3.44}$$

The input resistance r_π is now defined as $r_\pi=1/g_{be}$. Since $\beta_F=I_C/I_B$,

$$r_\pi=\frac{kT}{eI_B}=\frac{kT}{eI_C}\frac{I_C}{I_B}=\frac{\beta_F}{g_m} \tag{3.44a}$$

The capacitance C_μ is the small-signal capacitance of the collector-base space-charge region. The small-signal capacitance C_π comes about as follows. In a forward-biased *n–p–n* transistor electrons are injected into the base; an excess hole charge Q_B enters the base through the ohmic contact and is stored in the base to maintain space-charge neutrality. The definition of C_π is $C_\pi=\partial Q_B/\partial V_{BE}+C_{je}$, where C_{je} is the capacitance of the emitter-base space-charge region; for normal operation C_{je} is negligible. Since Q_B is then proportional to the collector current I_C, it follows that C_π is proportional to I_C.

We now show that $C_\pi r_\pi$ is proportional to β_F and independent of I_C. Since $r_\pi=\beta_F/g_m$,

$$\tau_\pi=C_\pi r_\pi=\frac{C_\pi}{g_m}\beta_F=\tau_0\beta_F \tag{3.44b}$$

so τ_π is proportional to β_F. In addition, the time constant $\tau_0=C_\pi/g_m$ is independent of I_C, since C_π and g_m are both proportional to I_C. Moreover, since I_C, hence g_m, is inversely proportional to the base width w_B, and Q_B, hence C_π, is proportional to w_B, τ_0 is proportional to the *square* of the base width. As a matter of fact, it may be shown that τ_0 is equal to the travel time τ_d of the electrons through the base region, given by Eq. (2.26).* Hence

$$\tau_0=\tau_d=\frac{w_B^{\ 2}}{2D_n} \tag{3.44c}$$

where D_n is the diffusion constant of electrons in the base region.

If the current I_C is quite small, the excess hole charge in the base region becomes negligible and the capacitance C_π equals the small-signal capacitance C_{je} of the emitter-base space-charge region, which is practically independent of current.

*A. van der Ziel, *Solid State Physical Electronics*, 3rd Ed., Prentice Hall, Englewood Cliffs, N.J., 1976, Chapter 16.

In order to write down the equations for the pulse response of the transistor, we must bear in mind that the Ebers–Moll equations are strongly nonlinear. It is therefore not convenient to use the base-emitter voltage as a variable. But we saw that the stored excess hole charge Q_B varies linearly with the current, hence seems to be a good parameter for describing the transient operation of transistors. This leads to the charge-control approach discussed in Section 3.4a.

3.4a Charge-Control Equations for Nonsaturated Transistors

We start with the equivalent circuit in Fig. 3.7 and assume that the transistor is an n–p–n transistor. The differential equation of the base circuit, neglecting the effects of the capacitance C_μ, is then

$$C_\pi \frac{dv_1}{dt} + \frac{v_1}{r_\pi} = i_b(t) \tag{3.45}$$

We now define

$$q_b(t) = C_\pi v_1(t) \tag{3.45a}$$

as the charge stored in the base. Introducing the time constant $\tau_\pi = C_\pi r_\pi$, we can write Eq. (3.45) as

$$\frac{dq_b}{dt} + \frac{q_b}{\tau_\pi} = i_b(t) \tag{3.45b}$$

since $v_1/r_\pi = C_\pi v_1/C_\pi r_\pi = q_b/\tau_\pi$

This is derived as a small-signal equation but, since τ_π is practically independent of the operating conditions, Eq. (3.45b) should be equally valid for large signals. This follows simply from steady-state considerations. Let Q_B be the dc charge stored in the base; then the recombination current is Q_B/τ_π, hence the dc base current $I_B = Q_B/\tau_\pi$, in accordance with Eq. (3.45b).

If applied to the large-signal pulse response, $q_b(t)$ is a positive charge. But for n–p–n transistors the minority carriers stored in the base are electrons which give a negative charge. However, this stored charge is compensated for by an equal and opposite hole charge, and it is this *positive* hole charge that makes up $q_b(t)$. We already have discussed this for the dc case.

According to Fig. 3.7,

$$i_c(t) = g_m v_1 = g_m r_\pi \frac{v_1}{r_\pi} = \beta_F \frac{q_b}{\tau_\pi} \tag{3.46}$$

since $g_m r_\pi = \beta_F$. This has the following simple interpretation: q_b/τ_π is the base recombination current and the collector current is β_F times as large. This again is derived as a small-signal equation but, since β_F and τ_π are independent of operating conditions, it also holds for large signals.

Equation (3.46) holds as long as the transistor is not saturated. In the circuit in Fig. 3.8 this is the case, if

$$i_c(t) < I_{Cs} \tag{3.46a}$$

where

$$I_{Cs} = \frac{V_{CC} - V_{CE,\text{sat}}}{R_C} \tag{3.46b}$$

We first assume that this condition is satisfied. If it is not satisfied, $i_c(t)$ will ultimately reach the value I_{Cs} and stay there.

Now let a voltage $V(t)$ be applied to the base by means of a resistor R_b (compare Fig. 3.8) and let $V(t) = V_1 > 0.70$ V for $t > 0$.

$$i_b(t) = I_{B1} = \frac{V_1 - 0.70}{R_b} \tag{3.47}$$

so Eq. (3.45) becomes

$$\frac{dq_b}{dt} + \frac{q_b}{\tau_\pi} = I_{B1} \tag{3.48}$$

with the initial condition $q_b(0^+) = 0$. In order to make sure that the

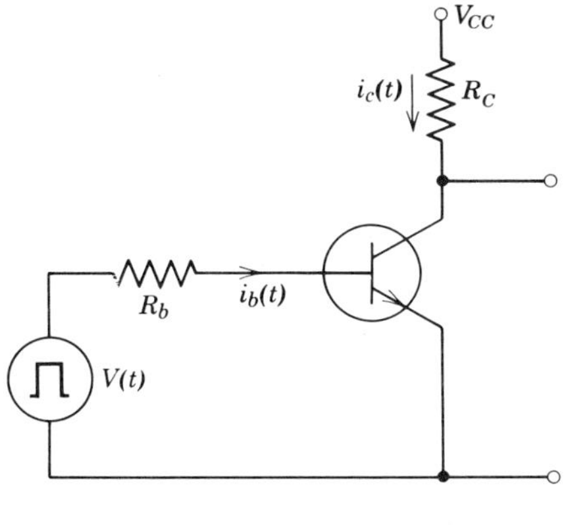

Figure 3.8. Large-signal transistor circuit.

transistor is never saturated, we require

$$\beta_F I_{B1} < I_{Cs} \quad \text{or} \quad I_{B1} < \frac{I_{Cs}}{\beta_F} = I_{Bs} \tag{3.48a}$$

We assume that this is the case.

To solve Eq. (3.48), we observe that $q_b(\infty) = I_{B1}\tau_\pi$, so the full solution is (compare Chapter 1)

$$q_b(t) = q_b(\infty)\left[1 - \exp\left(-\frac{t}{\tau_\pi}\right)\right] = I_{B1}\tau_\pi\left[1 - \exp\left(-\frac{t}{\tau_\pi}\right)\right] \tag{3.49}$$

or

$$i_c(t) = \beta_F I_{B1}\left[1 - \exp\left(-\frac{t}{\tau_\pi}\right)\right] \tag{3.50}$$

After a time $t_1 = 2.3\tau_\pi$, $i_c(t)$ reaches 90% of its final value.

Having evaluated the turn-on transient, we now investigate the turn-off transient. To this end we assume that $V(t)$ switches from $+V_1$ to $-V_2$ at $t=0$. Since at first there is still charge stored in the base, the base voltage remains at about 0.70 V, hence a current

$$i_b(t) = -I_{B2} = \frac{-V_2 - 0.70}{R_b} \tag{3.51}$$

flows into the base. Equation (3.45b) must now be written as

$$\frac{dq_b}{dt} + \frac{q_b}{\tau_\pi} = -I_{B2} \tag{3.52}$$

with the initial condition $q_b(0^+) = I_{B1}\tau_\pi$.

To solve Eq. (3.52) we observe that $q_b(\infty) = -I_{B2}\tau_\pi$; hence the full solution of Eq. (3.52) is

$$q_b(t) = q_b(0^+)\exp\left(-\frac{t}{\tau_\pi}\right) + q_b(\infty)\left[1 - \exp\left(-\frac{t}{\tau_\pi}\right)\right]$$

$$= I_{B1}\tau_\pi \exp\left(-\frac{t}{\tau_\pi}\right) - I_{B2}\tau_\pi\left[1 - \exp\left(-\frac{t}{\tau_\pi}\right)\right] \tag{3.53}$$

If $V_2 = 0.70$ V, $I_{B2} = 0$ and the last term in Eq. (3.53) disappears.

But $q_b(t)$ is a positive charge, hence the turn-off transient stops if

$q_b(t)=0$. Let this occur at $t=t_2$; then

$$(I_{B1}+I_{B2})\tau_\pi \exp\left(-\frac{t_2}{\tau_\pi}\right)-I_{B2}\tau_\pi=0$$

or

$$\exp\left(\frac{t_2}{\tau_\pi}\right)=\frac{I_{B1}+I_{B2}}{I_{B2}} \quad \text{or} \quad t_2=\tau_\pi \ln\left(\frac{I_{B1}+I_{B2}}{I_{B2}}\right) \tag{3.53a}$$

If $I_{B1}=I_{B2}$, we have

$$t_2=\tau_\pi \ln 2=0.69\tau_\pi \tag{3.53b}$$

If $V_2=0$ V, I_{B2} is much smaller than I_{B1} and t_2 is correspondingly larger. Switching from $+V_1$ to $-V_2$, rather than from $+V_1$ to 0 V, thus speeds up the response.

To be more exact, $V(t)$ should also switch from $-V_2$ to $+V_1$ for the turn-on transient. One then has to wait a certain time before $V_{BE}(t)$ reaches the value $V_{BE0}=0.70$ V needed to turn the transistor on. This is called the *turn-on delay*; it is discussed in Section 3.4f.

Example 1. If $C_\pi=10$ pF at $r_\pi=2000\ \Omega$, $R_b=5000\ \Omega$, $V_1=5.0$ V, $V_2=5.0$ V, and $V_{BE0}=0.70$ V, find (a) The time when the turn-on transient is 90% complete. (b) The full time for the turn-off response.

Answer. (a)

$$\tau_\pi=C_\pi r_\pi=2.0\times10^{-8}\ \text{s}$$

$$t_1=2.3\tau_\pi=4.6\times10^{-8}\ \text{s}$$

(b)

$$I_{B1}=\frac{V_1-0.70}{R_b}=\frac{5.0-0.70}{5000}=0.86\ \text{mA}$$

$$I_{B2}=\frac{5.0+0.70}{5000}=1.14\ \text{mA}$$

$$t_2=\tau_\pi \ln\left(\frac{I_{B1}+I_{B2}}{I_{B2}}\right)=2.0\times10^{-8}\ln\left(\frac{2.00}{1.14}\right)=1.12\times10^{-8}\ \text{s}$$

3.4b Influence of the Miller Effect on the Response

Next we take the effect of C_μ into account. Since C_μ gives rise to a Miller capacitance

$$(1+g_m R_c)C_\mu \cong g_m R_c C_\mu \tag{3.54}$$

in parallel with C_π, the total capacitance C_T in parallel with r_π is

$$C_T = C_\pi + g_m R_c C_\mu \tag{3.54a}$$

We now replace C_π in Eq. (3.45) by C_T, so

$$C_T \frac{dv_1}{dt} + \frac{v_1}{r_\pi} = i_b(t) \tag{3.55}$$

However, the stored charge is still as given by Eq. (3.45a), that is,

$$q_b(t) = C_\pi v_1$$

Therefore, switching to $q_b(t)$ as the new variable,

$$\frac{C_T}{C_\pi} \frac{dq_b}{dt} + \frac{q_b}{C_\pi r_\pi} = i_b(t) \tag{3.55a}$$

Introducing the time constants $\tau_\pi = C_\pi r_\pi$ and $\tau_T = C_T r_\pi = \tau_\pi + \beta_F R_c C_\mu$, we can write this equation as

$$\frac{dq_b}{dt} + \frac{q_b}{\tau_T} = \frac{\tau_\pi}{\tau_T} i_b(t) \tag{3.56}$$

with $i_b(t) = I_{B1}$ for the turn-on transient and $i_b(t) = -I_{B2}$ for the turn-off transient. Since β_F and τ_π are nearly independent of operating conditions, τ_T is also independent of operating conditions, so Eq. (3.56) can be considered a large-signal equation. Note that τ_T is proportional to β_F. Usually $\tau_T \cong 5$ to 10 times τ_π, so the Miller effect is *very* significant.

For the turn-on transient, if $I_{B1} < I_{Bs}$,

$$q_b(0^+) = 0 \qquad q_b(\infty) = I_{B1}\tau_\pi \tag{3.56a}$$

and, for the turn-off transient,

$$q_b(0^+) = I_{B1}\tau_\pi \qquad q_b(\infty) = -I_{B2}\tau_\pi \tag{3.56b}$$

As in the previous cases, we see that Eqs. (3.49) and (3.53) remain correct except that the expression $\exp(-t/\tau_\pi)$ must be replaced by $\exp(-t/\tau_T)$ everywhere. We thus have, for the turn-on transient,

$$q_b(t) = I_{B1}\tau_\pi \left[1 - \exp\left(-\frac{t}{\tau_T} \right) \right] \tag{3.57}$$

and, for the turn-off transient,

$$q_b(t) = I_{B1}\tau_\pi \exp\left(-\frac{t}{\tau_T}\right) - I_{B2}\tau_\pi\left[1 - \exp\left(-\frac{t}{\tau_T}\right)\right] \tag{3.58}$$

Since τ_T is proportional to β_F, and the transistor is never saturated ($I_{B1} < I_{BS}$), the transient becomes slower for transistors with a large β_F. It is therefore important that unsaturated transistors have values of β_F that are not too large. We shall see that this is less the case for saturated transistors.

One word of caution should be added here. The capacitance C_μ depends on the collector voltage V_{CE} and is thus a function of time. The above solution remains reasonably correct, however, if a suitable intermediate value is chosen for C_μ.

Example 2. Find by what factor the turn-on and turn-off transients are changed by the Miller effect if $\tau_\pi = 10^{-8}$ s, $\beta_F = 50$, $R_c = 2000\ \Omega$, and $C_\mu = 1.0$ pF.

Answer. $\tau_T = \tau_\pi + \beta_F R_c C_\mu = 10^{-8} + 50 \times 2 \times 10^3 \times 1.0 \times 10^{-12} = 11 \times 10^{-8}$ s, so the switching times are increased by a factor of $\tau_T/\tau_\pi = 11$.

3.4c Speedup of the Transient Response

It would be useful if the transient response of the circuit could be speeded up. It is easily seen from Fig. 3.9 how this can be achieved. One puts a capacitance C in parallel with the resistance R_b. If the effect of C_μ is first

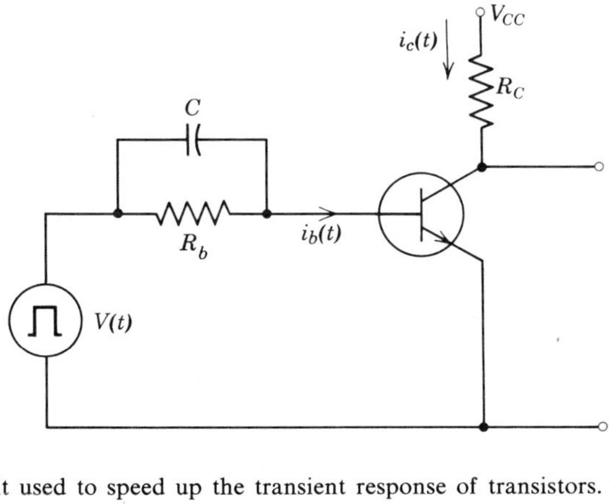

Figure 3.9. Circuit used to speed up the transient response of transistors. The pulse signal $V(t)$ is an emf with zero internal resistance.

ignored, one chooses C such that

$$CR_b = C_\pi r_\pi = \tau_\pi \tag{3.59}$$

because the circuit then acts as a simple voltage divider, transmitting the input step voltage directly to the base.

Example 3. If $R_b = 5000\ \Omega$ and $\tau_\pi = 2\times10^{-8}$ s, find the value of C needed to eliminate the turn-on response.

Answer.

$$C = \frac{\tau_\pi}{R_b} = \frac{2\times10^{-8}}{5000} = 4\ \text{pF}$$

If the effect of C_μ is taken into account, one must simply replace C_π by C_T. That is, Eq. (3.59) becomes

$$CR_b = C_T r_\pi = \tau_T \tag{3.59a}$$

In the approximation of Fig. 3.9, the transient is completely eliminated, and the response is instantaneous. This happens because the circuit is driven from an emf $V(t)$ of zero internal resistance. In a more realistic setting $V(t)$ comes from a previous stage and is developed across a resistance R_s. We then have the circuit in Fig. 3.10.

We now assume that I_1 and I_2 are large currents. Then the input impedance Z_π of the transistor is quite small and can be neglected in

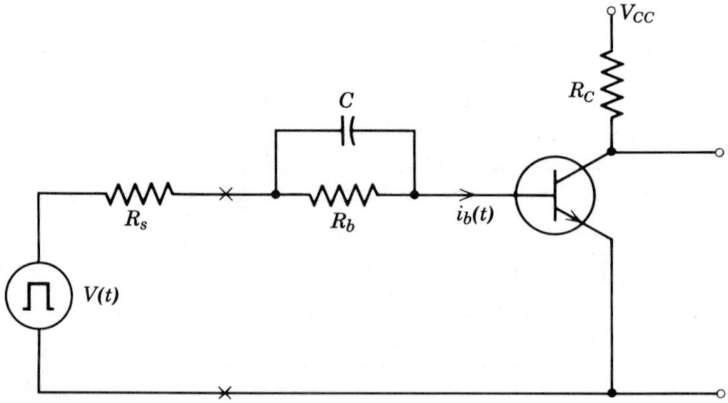

Figure 3.10. More realistic circuit, in which the internal resistance R_s of the pulse signal is taken into account.

comparison with R_s and R_b. Without speedup capacitor C the time constant of the transient is τ_T but, if C is chosen so that Eq. (3.59a) is satisfied, the time constant of the circuit is

$$\tau = \frac{CR_bR_s}{R_b + R_s} = \tau_T \frac{R_s}{R_b + R_s} \tag{3.60}$$

The transient is therefore no longer instantaneous but has been speeded up only by a factor of $(R_s + R_b)/R_s$.

3.4d Charge-Control Equations for Saturated Transistors

If, for the turn-on transient, $I_{B1} > I_{Bs}$, the collector current becomes saturated after some time. For the turn-off transient it takes the *storage delay time* before the collector current becomes unsaturated; after that the collector current goes rather rapidly to zero. We shall see that the unsaturated transient response is considerably speeded up, but that this is bought at the price of introducing the storage delay time.

The unsaturated part of the transient is easily evaluated. When the voltage $V(t)$ switches from 0 V to V_1, the differential equation of the turn-on transient becomes

$$\frac{dq_b}{dt} + \frac{q_b}{\tau_T} = I_{B1}\frac{\tau_\pi}{\tau_T} \tag{3.61}$$

with $q_b(0^+) = 0$. Since $q_b(\infty) = I_{B1}\tau_\pi$, the solution is

$$q_b(t) = I_{B1}\tau_\pi\left[1 - \exp\left(-\frac{t}{\tau_T}\right)\right] \tag{3.62}$$

as in the previous case. But when $q_b(t)$ reaches the value $I_{Bs}\tau_\pi$, $i_c(t)$ reaches the value I_{Cs} and the collector saturates. Let this occur at $t = t_1$; then

$$I_{Bs}\tau_\pi = I_{B1}\tau_\pi - I_{B1}\tau_\pi \exp\left(-\frac{t_1}{\tau_T}\right) \qquad \exp\left(-\frac{t_1}{\tau_T}\right) = \frac{I_{B1} - I_{Bs}}{I_{B1}}$$

or

$$t_1 = \tau_T \ln\left(\frac{I_{B1}}{I_{B1} - I_{Bs}}\right) = -\tau_T \ln\left(1 - \frac{I_{Bs}}{I_{B1}}\right) \tag{3.63}$$

Hence, if $I_{Bs} \ll I_{B1}$, since $-\ln(1-x) \cong x$ for small x,

$$t_1 \cong \tau_T \frac{I_{Bs}}{I_{B1}} \tag{3.63a}$$

Often, however, the full Eq. (3.63) must be used.

A considerable speedup of the response occurs. For example, if $I_{Bs} = 0.10 I_{B1}$, we have $t_1 \cong 0.1\tau_T$, whereas for the unsaturated transistor the transient is only 90% complete at $t_1 = 2.3\tau_T$.

We now investigate the turn-off transient. If $V(t)$ switches again from $+V_1$ to $-V_2$, it takes some time before the transistor becomes unsaturated. Let this occur at $t=0$; then, for $t>0$,

$$\frac{dq_b}{dt} + \frac{q_b}{\tau_T} = -I_{B2}\frac{\tau_\pi}{\tau_T} \tag{3.64}$$

with the initial condition $q_b(0^+) = I_{Bs}\tau_\pi$. Since $q_b(\infty) = -I_{B2}\tau_\pi$, the full solution to Eq. (3.64) is

$$q_b(t) = I_{Bs}\tau_\pi \exp\left(-\frac{t}{\tau_T}\right) - I_{B2}\tau_\pi\left[1 - \exp\left(-\frac{t}{\tau_T}\right)\right] \tag{3.65}$$

The transient stops when $q_b(t)=0$. Let this be at $t=t_2$; then

$$0 = (I_{B2} + I_{Bs})\tau_\pi \exp\left(-\frac{t_2}{\tau_T}\right) - I_{B2}\tau_\pi$$

or

$$\exp\left(\frac{t_2}{\tau_T}\right) = \frac{I_{B2} + I_{Bs}}{I_{B2}}$$

or

$$t_2 = \tau_T \ln\left(\frac{I_{B2} + I_{Bs}}{I_{B2}}\right) = \tau_T \ln\left(1 + \frac{I_{Bs}}{I_{B2}}\right) \tag{3.66}$$

so, for $I_{Bs} \ll I_{B2}$, since $\ln(1+x) \cong x$ for small x,

$$t_2 \cong \tau_T \frac{I_{Bs}}{I_{B2}} \tag{3.66a}$$

resulting again in a speedup of the response. Often, however, the full Eq. (3.66) must be used.

If we now look at the two equations

$$t_1 \cong \tau_T \frac{I_{Bs}}{I_{B1}} \qquad t_2 \cong \tau_T \frac{I_{Bs}}{I_{B2}} \tag{3.67}$$

and substitute $I_{Bs} = I_{Cs}/\beta_F$, we obtain

$$t_1 = \frac{\tau_T}{\beta_F}\frac{I_{Cs}}{I_{B1}} \qquad t_2 = \frac{\tau_T}{\beta_F}\frac{I_{Cs}}{I_{B2}} \tag{3.67a}$$

Here I_{Cs}, I_{B1}, and I_{B2} are determined by the external circuit, and the effect of the transistor itself is measured by the time constant

$$\frac{\tau_T}{\beta_F} = \frac{\tau_\pi}{\beta_F} + C_\mu R_c = \tau_0 + C_\mu R_c \tag{3.67b}$$

The properties of the transistor are contained in the parameters τ_0 and C_μ; the transistor should therefore be so designed that the base width is small and C_μ is small. Furthermore, the collector resistance R_c should not be chosen too large. Finally, the effect of β_F has largely disappeared.

Example 4. If $V_{CC} = 10$ V, $R_c = 2000\ \Omega$, $\beta_F = 50$, $V_{CE,\text{sat}} \cong 0$ V, $R_b = 5000\ \Omega$, $V_1 = 5.0$ V, $V_2 = 5.0$ V, and $\tau_T = 30 \times 10^{-8}$ s, find t_1 and t_2.

Answer. As in Example 1, $I_{B1} = 0.86$ mA and $I_{B2} = 1.14$ mA. According to Eq. (3.46b),

$$I_{Cs} = \frac{10}{2000} = 5\text{ mA} \qquad I_{Bs} = \frac{I_{Cs}}{50} = 0.10\text{ mA}$$

$$t_1 = \tau_T \ln\left(\frac{I_{B1}}{I_{B1} - I_{Bs}}\right) = 30 \times 10^{-8} \ln\left(\frac{0.86}{0.76}\right) = 3.7 \times 10^{-8}\text{ s}$$

$$t_2 = \tau_T \ln\left(\frac{I_{B2} + I_{Bs}}{I_{B2}}\right) = 30 \times 10^{-8} \ln\left(\frac{1.24}{1.14}\right) = 2.5 \times 10^{-8}\text{ s}$$

Having calculated the turn-on transient and the turn-off transient, we evaluate the turn-off delay in Section 3.4e.

3.4e Response during Saturation

We now investigate the response during saturation. Equations (3.48) and (3.52) are not valid, but we can write a similar equation for the excess stored charge, $q_s(t) = q_b(t) - I_{Bs}\tau_\pi$, if the following points are taken into account.

Only the part I_{Bs} of the injected current I_{B1} will be used to keep the transistor saturated. The *excess current* $I_{B1} - I_{Bs}$ will be used in two ways:

1. To increase the excess stored charge q_s. In analogy with the previous case this occurs at the rate dq_s/dt.

2. To maintain the excess recombination current. In analogy with the previous case we write this term as q_s/τ_s, where τ_s is the time constant associated with the excess charge.

The turn-on transient may thus be described by the equation

$$\frac{dq_s}{dt} + \frac{q_s}{\tau_s} = I_{B1} - I_{Bs} \tag{3.68}$$

where τ_s is defined as

$$\tau_s = \left.\frac{q_s}{I_{B1} - I_{Bs}}\right|_{\text{steady state}} \tag{3.68a}$$

The time constant τ_s may be different from τ_π, because recombination in a saturated and in a nonsaturated transistor may occur in different parts of the device. Calculation* gives

$$\tau_s = \frac{(1+\beta_F)\tau_R + (1+\beta_R)\tau_\pi}{1+\beta_F+\beta_R} \tag{3.68b}$$

where β_F and β_R are the forward and the reverse current amplification factors, respectively, defined as $\beta_F = \alpha_F/(1-\alpha_F)$ and $\beta_R = \alpha_R/(1-\alpha_R)$. The time constant τ_π is associated with the forward-biased transistor, and the time constant τ_R is associated with the reverse operation.

A symmetrical transistor, that is, a transistor for which the characteristic does not change if the emitter and the collector are interchanged, has generally $\beta_F = \beta_R$ and $\tau_\pi = \tau_R$; hence $\tau_s = \tau_\pi$. For an asymmetrical transistor

*A. van der Ziel, *Solid State Physical Electronics*, 3rd Ed. Prentice Hall, Englewood Cliffs, N.J., 1976, Chapter 17.

$\beta_F \gg \beta_R$, and in that case $\tau_s \cong \tau_R$. If the forward-biased collector gives rise to a large injection of holes from the base into the collector, τ_R may be quite large.

We now return to the solution to Eq. (3.68). Let the saturated part of the turn-on transient start at $t=0$; then $q_s(0^+)=0$. Since $q_s(\infty)=(I_{B1}-I_{Bs})\tau_s$, the solution to Eq. (3.68) is

$$q_s(t)=(I_{B1}-I_{Bs})\tau_s\left[1-\exp\left(-\frac{t}{\tau_s}\right)\right] \tag{3.69}$$

and it takes a time of $2.3\tau_s$ before the stored charge reaches 90% of its full value. The transistor remains saturated all the time, so $i_c(t)$ does not change and the transient does not show up in the collector response.

For the turn-off transient we have under saturated conditions if the turn-off starts at $t=0$:

$$\frac{dq_s}{dt}+\frac{q_s}{\tau_s}=-I_{B2}-I_{Bs} \tag{3.70}$$

with $q_s(0^+)=(I_{B1}-I_{Bs})\tau_s$. Since $q_s(\infty)=-(I_{B2}+I_{Bs})\tau_s$, the solution to Eq. (3.70) is

$$q_s(t)=(I_{B1}-I_{Bs})\tau_s\exp\left(-\frac{t}{\tau_s}\right)-(I_{B2}+I_{Bs})\tau_s\left[1-\exp\left(-\frac{t}{\tau_s}\right)\right] \tag{3.71}$$

This part of the transient stops if the excess stored charge $q_s(t)$ becomes zero. Let this be at $t=t_3$; then

$$(I_{B1}+I_{B2})\tau_s\exp\left(-\frac{t_3}{\tau_s}\right)-(I_{B2}+I_{Bs})\tau_s=0$$

or

$$\exp\left(\frac{t_3}{\tau_s}\right)=\frac{I_{B2}+I_{B1}}{I_{B2}+I_{Bs}} \qquad t_3=\tau_s\ln\left(\frac{I_{B2}+I_{B1}}{I_{B2}+I_{Bs}}\right) \tag{3.72}$$

This is the *storage delay time* of the circuit.

If $I_{B2}\gg I_{Bs}$ and $I_{B1}=I_{B2}$, this results in

$$t_3\cong\tau_s\ln 2=0.69\tau_s \tag{3.72a}$$

Usually, however, $V_2=0$, and then I_{B2} is considerably smaller than I_{B1} and

Eq. (3.72) must be used. According to Eq. (3.72), this slows down the response.

Example 5. If $\tau_s = 2\times10^{-8}$ s, find t_3 if $I_{B1} = I_{B2} \gg I_{Bs}$.

Answer. $t_3 = 0.69\tau_s = 1.4\times10^{-8}$ s.

3.4f Turn-on Delay in Transistors

If the base voltage $V(t)$ of a transistor is switched from $-V_2$ to $+V_1$ at $t=0$, it takes some time before V_{BE} reaches the value $V_{BE0} = 0.70$ V needed to turn the transistor on. This time is called the *turn-on delay time*.

We should be somewhat cautious about the value of V_{BE} for which the transistor draws current. If the emitter current is 1.0 mA at $V_{BE} = V_{BE0} = 0.70$ V, it may be assumed that the transistor begins to draw current for $V_{BE} > V_{BE1} \cong 0.55$ V. The stored charge may thus be neglected for $V_{BE} < V_{BE1}$ but must be taken into account for $V_{BE1} \leqslant V_{BE} \leqslant V_{BE0}$.

The equivalent circuit for this case is shown in Fig. 3.11, where C_{in} is the input capacitance of the transistor.

$$C_{\text{in}} = C_{je} + C_{\mu} \tag{3.73}$$

Here C_{je} is the differential capacitance of the emitter-base space-charge region* and C_{μ} is the differential capacitance of the collector-base space-charge region. Both capacitances depend on the applied voltage; to simplify calculations, an appropriately chosen effective value for C_{in} may be used. We assume that this has been done. We then have for $t > 0$,

$$C_{\text{in}}\frac{dV_{BE}}{dt} = \frac{V_1 - V_{BE}}{R_b} \qquad \frac{dV_{BE}}{dt} + \frac{V_{BE}}{\tau_b} = \frac{V_1}{\tau_b} \tag{3.74}$$

with $\tau_b = C_{\text{in}}R_b$. The initial condition is $V_{BE}(0^+) = -V_2$, whereas $V_{BE}(\infty) = V_1$; according to Eq. (3.74), we thus have

$$V_{BE} = -V_2\exp\left(-\frac{t}{\tau_b}\right) + V_1\left[1 - \exp\left(-\frac{t}{\tau_b}\right)\right] \tag{3.75}$$

We now substitute $V_{BE}(t_1) = V_{BE1} = 0.55$ V and calculate the turn-on delay time t_0. We assume here that for $-V_2 < V_{BE} < V_{BE1} = 0.55$ V the capaci-

*Because the transistor is off, the stored charge in the base is zero, or $C_{\pi} = C_{je}$, as mentioned earlier.

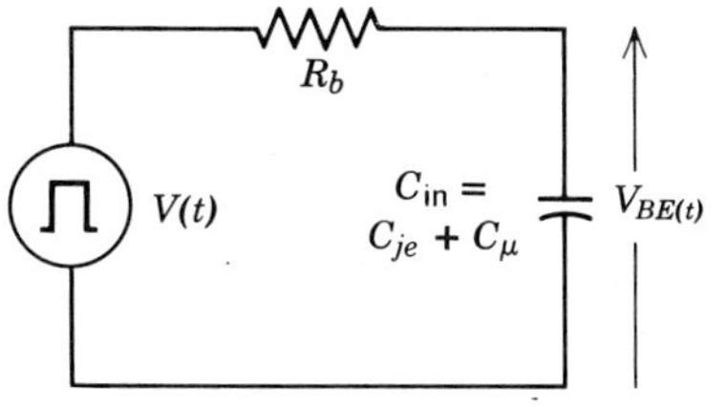

Figure 3.11. Equivalent circuit of the input of the transistor for the case in which the device does not draw current.

tance C_{in} is charged, whereas for $V_{BE1} < V_{BE} < V_{BE0} = 0.70$ V the stored charge in the base builds up. This yields

$$V_{BE1} = V_1 - (V_1 + V_2)\exp\left(-\frac{t_0}{\tau_b}\right) \quad \text{or} \quad \exp\left(\frac{t_0}{\tau_b}\right) = \frac{V_1 + V_2}{V_1 - V_{BE1}}$$

or

$$t_0 = \tau_b \ln\left(\frac{V_1 + V_2}{V_1 - V_{BE1}}\right) \tag{3.75a}$$

Example 6. Find t_0 if $C_{in} = 2.0$ pF, $R_b = 5000\ \Omega$, $V_1 = V_2 = 5.0$ V, and $V_{BE1} = 0.55$ V.

Answer. $\tau_b = C_{in}R_b = 2 \times 10^{-12} \times 5 \times 10^3 = 1.0 \times 10^{-8}$ s. $t_0 = 1.0 \times 10^{-8} \ln(10/4.45) = 1.0 \times 10^{-8} \times 0.81 = 0.81 \times 10^{-8}$ s.

For the turn-off transient we saw that the transient in $q_b(t)$, hence in $i_c(t)$, stopped at the time at which $q_b(t) = 0$. As far as $i_c(t)$ is concerned, the transient is then complete. But it takes a certain time, the *recovery time*, before $V_{BE}(t)$ drops from the value $V_{BE1} = 0.55$ V, where the current ceases to flow, to the value $-V_2$. This recovery time t_4 can be calculated similarly to the turn-on delay t_0. Equation (3.74) becomes

$$\frac{dV_{BE}}{dt} + \frac{V_{BE}}{\tau_b} = -\frac{V_2}{\tau_b} \tag{3.76}$$

with $V_{BE} = V_{BE1}$ at $t = 0^+$. This yields $t_4 \cong 2.3\tau_b$ for 90% of the full change in response.

Having calculated the various parts of the transient response in saturated transistors, we now present a schematic picture in Fig. 3.12, showing the turn-on delay t_0, the turn-on time t_1, the turn-off delay t_3, and the turn-off time t_2.

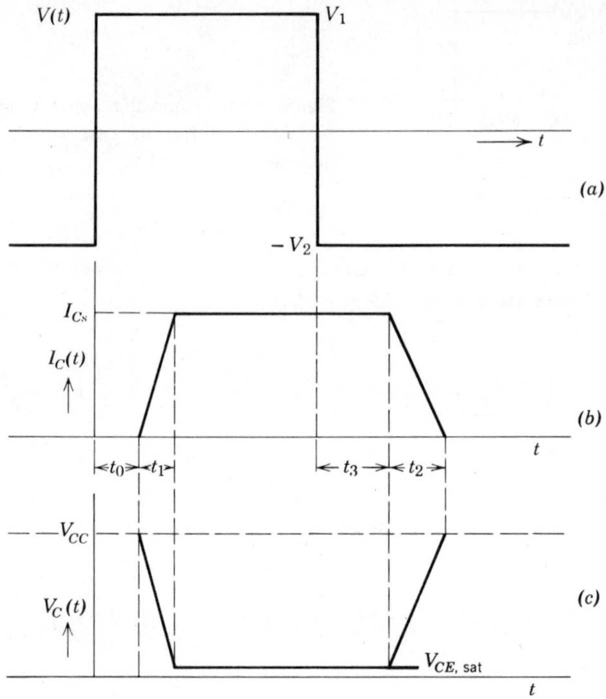

Figure 3.12. Pulse response of circuit for a saturated transistor. (*a*) Applied input pulse $V(t)$. (*b*) Resulting collector current $I_C(t)$ versus time. (*c*) Resulting collector voltage $V_C(t)$ versus time. I_{Cs} is the collector saturated current; $V_{CE,\mathrm{sat}}$ is the collector saturated voltage.

Example 7. Combining the results from Examples 4, 5, and 6, find the total turn-on time and the total turn-off time.

Answer. $t_{on}=t_0+t_1=0.8\times10^{-8}+3.7\times10^{-8}=4.5\times10^{-8}$ s. $t_{off}=t_3+t_2=1.4\times10^{-8}+2.5\times10^{-8}=3.9\times10^{-8}$ s.

In our examples we assumed $-V_2=-5$ V. Usually, however, $V_2=0$; this results in an increase in the turn-off delay t_3 and the turn-off time t_2, since I_2 is much smaller in these cases, whereas the turn-on delay t_0 is much shorter.

The recovery time t_4 is important for determining the repetition rate of the pulses.

3.5 THE EARLY EFFECT

If for a transistor the collector current I_C is plotted versus V_{CE} for constant base current I_B, one obtains a set of straight lines for $V_{CE}>0$ which meet the V_{CE} axis at the same point $-V_{CE0}$ (Fig. 3.13). This is called the *Early effect*, and $-V_{CE0}$ is called the *Early voltage*. For large base currents I_B, such as occur in transistor pulse circuits, the slope $h_{0e}=(\partial I_C/\partial V_{CE})_{I_B=\text{const}}$ can be quite large. This has two significant effects:

1. The Miller effect is reduced, since the ac resistance $1/h_{0e}$ is in parallel with the collector resistance R_c. Hence

$$\tau_T=\tau_\pi+\beta_F C_\mu \frac{R_c}{1+h_{0e}R_c} \tag{3.77}$$

This speeds up the response to a certain extent.

2. In pulse circuits with a capacitive load C_0 the expression for the time t_1 needed for the discharging of C_0 during the turn-on response is changed to

$$t_1=\frac{C_0\left(V_{CE}-V_{CE,\text{sat}}\right)}{\bar{I}_C} \tag{3.78}$$

where $\bar{I}_C$ is the average value of I_C. This parameter must be introduced, since I_C depends on V_{CE}, hence on time.

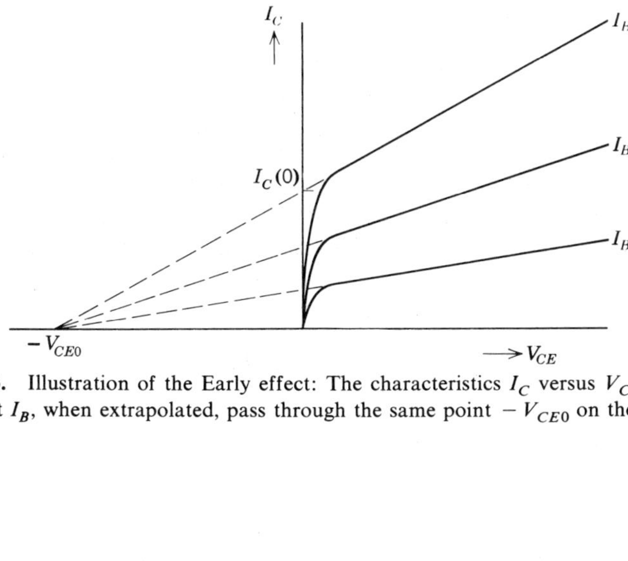

Figure 3.13. Illustration of the Early effect: The characteristics I_C versus V_{CC} at constant base current I_B, when extrapolated, pass through the same point $-V_{CE0}$ on the V_{CE} axis.

If $I_C(0)$ is the collector current extrapolated to $V_{CE}=0$, then, because of Early effect,

$$I_C = I_C(0)\left(1+\frac{V_{CE}}{V_{CE0}}\right) \quad \text{or} \quad \bar{I}_C = I_C(0)\left(1+\frac{1}{2}\frac{V_{CC}}{V_{CE0}}\right) \tag{3.78a}$$

since V_{CE} changes linearly from V_{CC} at $t=0$ to $V_{CE,\text{sat}}(\cong 0)$ at $t=t_1$.

PROBLEMS

3.1 In an n–p–n transistor with forward-biased collector junction the emitter is left floating, that is, $I_E=0$. Show that $V_{EC}=-(kT/e)\ln\alpha_F$.

3.2 In a symmetrical transistor, with $\alpha_F=\alpha_R=0.95$ and $I_{ES}=I_{CS}$, the emitter and collector are connected to make a diode with the base contact as the other electrode. When the transistor is connected for normal bias, $I_E=1.0$ mA at $V_{BE}=V_{BE0}=0.70$ V. Find the value of V_{BE} for a diode current of 1.0 mA.

3.3 In an n–p–n transistor $\alpha_F=0.995$ and $\alpha_R=0.50$. The device is made into a diode by letting the collector float. When the transistor is connected as a normal forward-biased transistor, $V_{BE}=0.700$ V at $I_E=1.00$ mA. Find the value of V_{BE} for a diode current of 1.0 mA.

3.4 An n–p–n transistor has $\beta_F=100$ and $\alpha_R=0.10$. Plot V_{CE} versus I_B/I_C at $T=300°$K with the help of Eq. (3.16). For what value of I_B/I_C is $V_{CE}=0.20$ V?

3.5 In the circuit in Fig. 3.3a, $R_b=2000\ \Omega$, $R_c=1000\ \Omega$, $V_{CC}=5.0$ V, $V_{BE0}=0.70$ V, $V_{CE,\text{sat}}=0.10$ V and $\beta_F=25$. Find V'_{BB}.

3.6 In the circuit in Fig. 3.14 evaluate V_E as a function of V_{BB}.

(a) When the transistor is not saturated ($V_C-V_E \geqslant V_{CE,\text{sat}}$).
(b) When the transistor is saturated ($V_C-V_E=V_{CE,\text{sat}}$).
(c) Find the value of V_{BB} for which the transistor is just saturated.
Hint. A transistor is just saturated if $I_C=\beta_F I_B$ and $V_{CE}=V_{CE,\text{sat}}$. Take $V_{BE0}=0.70$ V; $V_{CE,\text{sat}}=0.10$ V; $\beta_F=50$.

3.7 Find the equivalent circuit for Fig. 3.14.

(a) When the transistor is not saturated.
(b) When the transistor is saturated.
(c) Explain the difference.

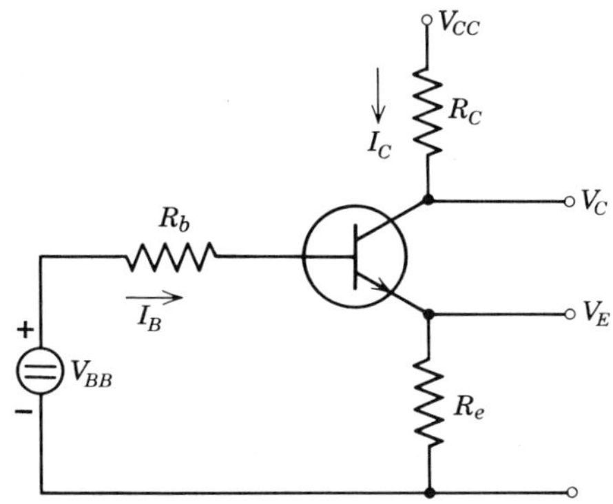

Figure 3.14

3.8 In the circuit in Fig. 3.5, $\beta_F = 50$, $R_b = 2000\ \Omega$, $R_c = 1000\ \Omega$, $V_{CC} = 4.0$ V, $V(t) = 4.0$ V, $C_0 = 50$ pF, $V_{BE0} = 0.70$ V, and $V_{CE,\text{sat}} = 0.00$ V.

(a) Calculate the turn-on time accurately.
(b) Compare the results from (a) with the approximate formula of Eq. (3.36a).
(c) Calculate the turn-off time (to 90% of the final value).

3.9 The circuit in Fig. 3.15 is used to decrease the turn-off time of a transistor. Calculate the turn-off time and compare it with the value obtained when the diode D is not present. $R_c = 1000\ \Omega$, $C_0 = 50$ pF, $\beta_F = 50$, $R_b = 2000\ \Omega$, $V_{CE,\text{sat}} = 0.00$ V, and the turn-on voltage of the diode $D = 0.70$ V. *Hint*. Define the turn-off time as the time it takes V_0 to rise from 0.0 V to its final value, and the turn-on time as the time it takes V_0 to drop to 0 V. When D is not present use the 90% response time for turn-off.

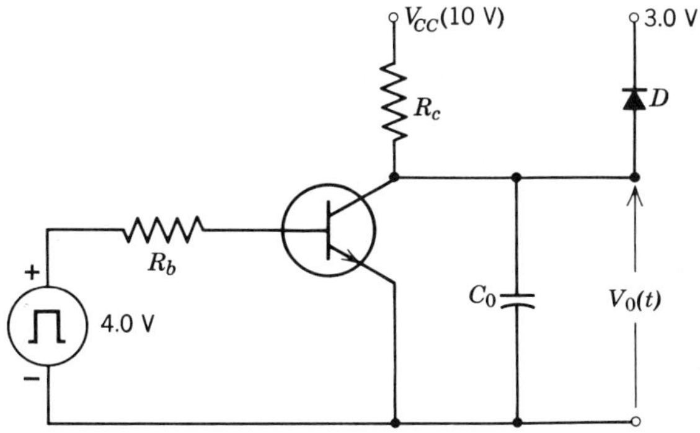

Figure 3.15

3.10 In the circuit in Fig. 3.6a, $R_b = 2000\ \Omega$, $R_e = 1000\ \Omega$, $C_0 = 50$ pF, and $\beta_F = 100$. Find the turn-on and turn-off time, defined as the time it takes to obtain 90% of the full response.

3.11 The circuit in Fig. 3.16 is sometimes used to decouple the voltage supply from the circuit. Find V_0 if $C = 0.1\ \mu$F and the voltage $V(t)$ is a square wave with a 4-V amplitude and a 50-KHz frequency. Assume that the transistor is saturated when on. $V_{CE,\text{sat}} = 0.00$ V and $V_{BE0} = 0.70$ V. *Hint*. When the transistor is off, the capacitance C is charged through 1000 Ω. When the transistor is on, the capacitance C is discharged. A stable equilibrium situation is obtained if the average charging current equals the average discharging current. This helps one to find V_C.

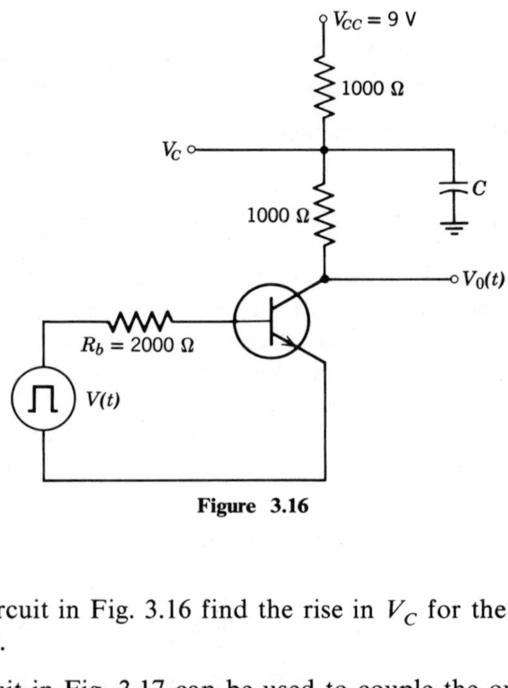

Figure 3.16

3.12 In the circuit in Fig. 3.16 find the rise in V_C for the half period the transistor is off.

3.13 The circuit in Fig. 3.17 can be used to couple the output pulse of a transistor pulse circuit into a load. $V(t)$ is a 4-V-amplitude square wave of 50-KHz frequency. Find $V_0(t)$. Assume that the transistor is saturated when on. $V_{CE,\text{sat}} = 0.00$ V and $V_{BE0} = 0.70$ V. *Hint*. Assume a value of the capacitor voltage V_C and find V_C by requiring that the charging current during the off period equal the discharging current during the on period. Bear in mind that $V_{CE} = 0$ during the on period.

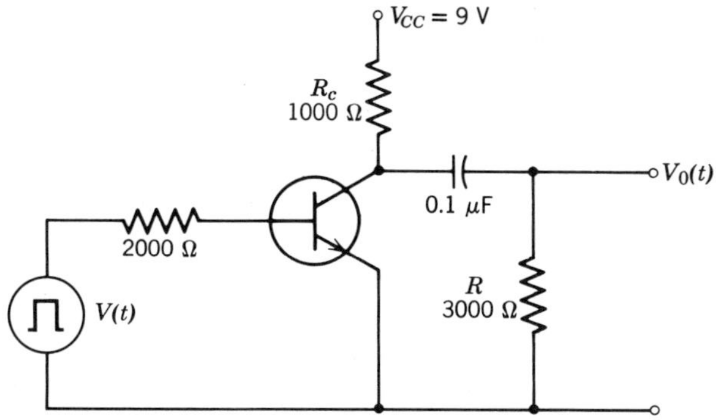

Figure 3.17

3.14 How much does V_C change during the discharge cycle in Problem 3.13?

3.15 In the circuit in Fig. 3.17 the capacitor and load resistor in Problem 3.13 are replaced by an emitter follower. $\beta_F = 100$ and $R_E = 1000\ \Omega$ (Fig. 3.18).

(a) Find $V_0(t)$ if $V_{BE0} = 0.70$ V.
(b) Compare with the result of Problem 3.13 and decide which circuit is more suitable.

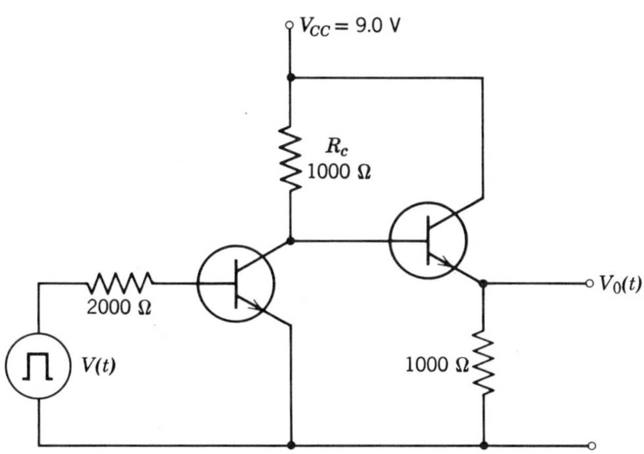

Figure 3.18

3.16 A transistor has β_F ranging between 30 and 100. All transistors have $C_\pi = 15$ pF at $I_C = 1$ mA, and $C_\mu = 1.5$ pF. $R_c = 1000\ \Omega$ and $C_\pi r_\pi$ is independent of current. Find the range of values for τ_π and τ_T. What do you conclude from your result?

3.17 The input time constant τ_T of a transistor, including the Miller effect, is 10^{-7} s. If the base resistance R_b is 2000 Ω, find the capacitance C needed to remove the turn-on transient.

3.18 In Problem 3.17 the base is fed from a source $V(t)$ of 1000-Ω internal resistance. If C is chosen as in Problem 3.17, find the time constant of the input circuit.

3.19 In a switching transistor the input voltage switches from −5.0 V to +5.0 V and back. ($V_1 = V_2 = 5.0$ V.)

(a) Find the turn-on delay of the transistor.
(b) Find the turn-on time.
(c) Find the turn-off delay.
(d) Find the turn-off time.
(e) Evaluate the total time for the turn-on and for the turn-off. $V_{BE0}=0.70$ V, $V_{BE1}=0.55$ V, $\tau_\pi = 2\times10^{-8}$ s, $\tau_s = 2\times10^{-8}$ s = time constant for excess carriers, $C_\mu = 1.0$ pF, $\beta_F = 50$, $R_c = 2000$ Ω, $R_b = 4300$ Ω, $V_{CC} = 8.0$ V, $V_{BE} = 0.70$ V, $V_{CE,\mathrm{sat}} \cong 0.00$ V, and the input capacitance C_{in} for devices when off = 2.0 pF.

3.20 Repeat Problem 3.19 if the input voltage switches from 0 V to 5.0 V ($V_1 = 5.0$ V, $V_2 = 0$) and back, and discuss the differences.

3.21 The buffer circuit shown in Fig. 3.19 drives a 200-pF capacitive load C. The diode has $V_{D0} = 0.70$ V, and the *n–p–n* transistors have $V_{BE0} = 0.70$ V, $\beta_F = 50$, $V_{CE,\mathrm{sat}} = 0.15$ V. The maximum current available from the collector supply is 10 mA under steady-state conditions (Fig. 3.19*a*).

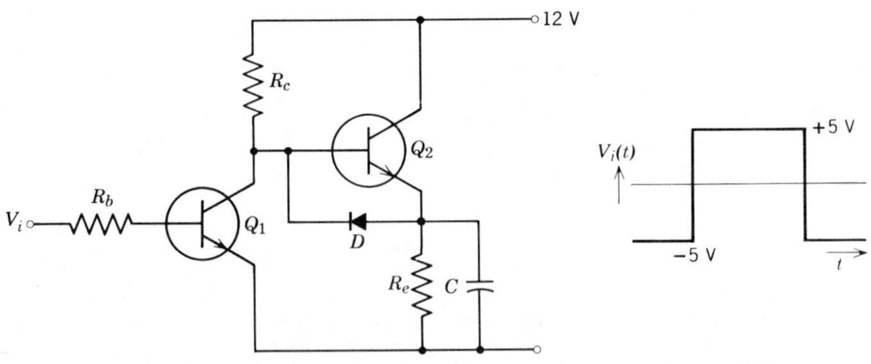

Figure 3.19

(a) Find the values at R_b, R_c, and R_e if Q_1 is to be overdriven into saturation by a factor of 3 using the input waveform v_i shown.
(b) Estimate the rise and fall times of the output $v_2(t)$ (Fig. 3.19*b*).
(c) Sketch $v_2(t)$. Label the amplitudes and time intervals.
(d) Estimate the storage delay time for Q_1 if $\tau_s = 10$ ns.

CHAPTER

4 NONLINEAR FET MODELS AND CIRCUITS

In many applications FETs are used as nonlinear devices, usually in one of two modes of operation:

1. The device is off; that is, $I_D = 0$.

2. The device is operating in the ohmic regime, that is, the drain voltage is close to zero and the output behaves as a voltage-controlled resistor.

To discuss this, we need a review of FET operation, which is given in Section 4.1. Section 4.2 discusses FETs as off-on switches. Section 4.3 deals with an integrated metal oxide semiconductor field-effect transistor (MOSFET) circuit without resistors. Section 4.4 describes the pulse response of FETs, first the intrinsic response, and then the response when the circuit is operating into a capacitive load. Then the complementary symmetry MOSFET circuit (CMOS or COSMOS) is considered; it has the interesting features that it draws current only during switching, that switching is faster, and that no resistors are needed in the circuit. Next a MOSFET circuit with a depletion mode load is discussed. Section 4.5 deals with device capacitances and their use in memory circuits.

4.1 REVIEW OF FET OPERATION

In n-channel junction FETs (JFETs) the characteristic of the device for $V_{DS} \geqslant 0$, $V_{GS} \geqslant V_P$, can be approximated as

$$I_D = K\left[2(V_{GS} - V_P)V_{DS} - V_{DS}^2\right] \tag{4.1}$$

for $V_{DS} \leqslant V_{GS} - V_p$ (non-pinch-off regime) and

$$I_D = K(V_{GS} - V_p)^2 \tag{4.2}$$

for $V_{DS} \geqslant V_{GS} - V_p$ (pinch-off regime). If V_{DS} is close to zero, the V_{DS}^2 term in Eq. (4.1) can be neglected, and Eq. (4.1) can be written

$$I_D = 2K(V_{GS} - V_p)V_{DS} = \frac{V_{DS}}{R_{\text{on}}} \tag{4.3}$$

where $R_{\text{on}} = [2K(V_{GS} - V_p)]^{-1}$ is the output resistance of the device, hence the term "ohmic regime." For $V_{GS} \leqslant V_P$, $I_D = 0$.

In these equations K is a constant of the order of 1 mA/V^2, V_{GS} is the gate-source voltage, V_{DS} (>0) is the drain-source voltage, and V_p (<0) is the pinch-off voltage, since $I_D = 0$ for $V_{GS} < -V_p$. To prevent the flow of gate current, V_{GS} should be kept smaller than the turn-on voltage of the gate-source diode ($\cong 0.70$ V in silicon devices). If the gate voltage can run positive, it is advisable to protect the gate with a series resistor, so V_{GS} will hardly exceed the turn-on voltage of the gate-source diode and the gate current will be limited. V_{DS} should not be made too large, to prevent avalanche breakdown in the drain region. For normal operation $V_p \leqslant V_{GS} \leqslant 0.70$ V, but for the off condition $V_{GS} < V_p$.

If the conducting channel has a width w and a length L, it can be shown that

$$K = K_0 \frac{w}{L} \tag{4.1a}$$

where K_0 is a constant independent of the two dimensions.

In the p-channel JFET all polarities are opposite and all currents flow in opposite directions. We thus have, instead of Eqs. (4.1) through (4.3),

$$I_D = -K\left[2(V_{GS} - V_p)V_{DS} - V_{DS}^2\right] \tag{4.4}$$

for $V_{DS} \geqslant V_{GS} - V_p$ (non-pinch off regime) and

$$I_D = -K(V_{GS} - V_p)^2 \tag{4.5}$$

for $V_{DS} \leqslant V_{GS} - V_p$ (pinch-off regime), whereas in the ohmic regime

$$I_D = -2K(V_{GS} - V_p)V_{DS} = \frac{V_{DS}}{R_{\text{on}}} \tag{4.6}$$

where $R_{\text{on}} = [-2K(V_{GS} - V_p)]^{-1}$. In this case $V_p > 0$, and $V_{DS} < 0$. We must also choose $V_{GS} \geqslant -0.70$ V to prevent gate current, so $V_p \geqslant V_{GS} \geqslant -0.70$ V. For $V_{GS} \geqslant V_P$, $I_D = 0$.

In the n-channel MOSFET Eqs. (4.1) through (4.3) are valid, but it is common practice to replace the pinch-off voltage V_p by the turn-on voltage V_T. Since the gate is insulated from the channel, there is no restriction on V_{GS}, and no gate current flows for any reasonable gate bias. There are two possibilities:

1. $V_T < 0$ and $V_{GS} \geqslant V_T$. Such devices are called *depletion-mode* devices; there is a conducting channel at $V_{GS} = 0$.

2. $V_T > 0$ and $V_{GS} \geqslant V_T$. Such devices are called *enhancement-mode* devices; there is no conducting channel at $V_{GS} = 0$.

Many n-channel MOSFETs are depletion-mode devices, but by proper design one can make devices that operate in the enhancement mode only.

In the p-channel MOSFET Eqs. (4.4) through (4.6) are valid, but it is also common practice to replace the pinch-off voltage V_p by the turn-on voltage V_T. Again no gate current flows for any reasonable gate bias, so there is no restriction on V_{GS}. Furthermore, all present p-channel MOSFETs have $V_T < 0$; since $V_{GS} < V_T$ in order to have a conducting channel, these devices always operate in the enhancement mode.

We shall see that enhancement-mode devices have real advantages in switching and logic circuits.

It is common practice to picture FETs as three-terminal devices with gate, source, and drain electrodes. Figure 4.1*a* and *b* show the symbols for an n-channel and a p-channel JFET, respectively; for an n-channel device the arrow points toward the gate, and for a p-channel device the arrow points in the other direction. Sometimes one distinguishes between the source and the drain by drawing the gate electrode closest to the source electrode. Figure 4.1*c* and *d* show the symbols for an n-channel and a p-channel MOSFET, respectively. The current flows *out of* the source for an n-channel device and *into* the source for a p-channel device, so the symbol automatically designates the source electrode. The symbols do not discriminate between the enhancement and the depletion mode.

MOSFETs have a fourth electrode: the substrate of the device. Many textbooks include this fourth electrode, thereby cluttering up the circuit diagrams. For the sake of clarity we do not do this.

The bias of the substrate affects the device operation. The reason is that there is a space-charge region between the conducting channel and the substrate, the width of which depends on the gate and the substrate bias. For an n-channel device the width of the space-charge region increases when the substrate is biased more negatively with respect to the source.

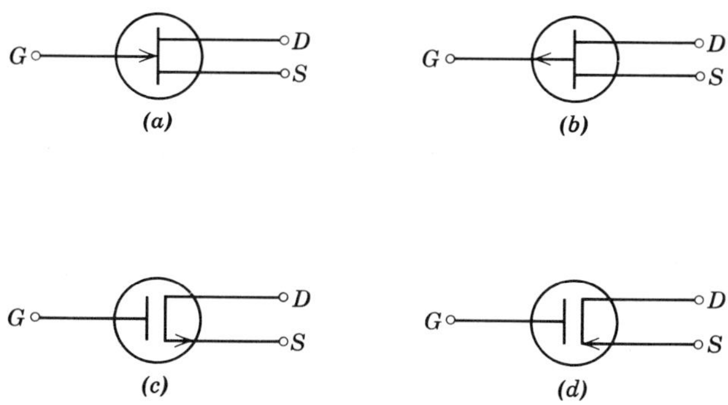

Figure 4.1. Symbols for JFETs and MOSFETs. (*a*) Symbol for *n*-channel JFET. (*b*) Symbol for *p*-channel JFET. (*c*) Symbol for *n*-channel MOSFET. (*d*) Symbol for *p*-channel MOSFET.

This decreases the mobile charge available in the channel and therefore raises the turn-on voltage V_T of the channel. Proper biasing of the substrate is therefore important.

Unless otherwise stated, we assume that the substrate is connected to the source electrode or kept at source potential.

4.2 FETS AS OFF-ON SWITCHES

FETs used in logic circuits are either *off* (carrying zero current) or *on* (carrying current), and in the on condition the drain voltage is so close to zero that the output acts as a resistor of value R_{on}. We now discuss the applications in greater detail.

4.2a *n*-Channel JFET

Let the *n*-channel JFET have a pinch-off voltage $V_p < 0$ and let the gate circuit have a protecting resistor R_g in the gate lead to limit the gate current (Fig. 4.2*a*).

The device is off if $V_i < V_p < 0$, and in this case $V_0 = V_{DD}$. To bring the device to the on condition, a sufficiently large positive voltage V_i is applied, driving the gate into the gate current regime. The gate voltage V_{GS} is then approximately equal to the turn-on voltage of a silicon diode, or about 0.70 V; this value is called V_{GS0}. The condition for V_i is therefore $V_i > V_{GS0}$. We limit use of the symbol V_{GS0} to JFETs in which the gate draws current limited by the gate resistance R_g in the circuit.

We now evaluate the transfer characteristic of the circuit in Fig. 4.2*a*. For $V_{GS} < V_p$, $V_0 = V_{DD}$, and the device current is zero. For $V_{GS} > V_p$ and

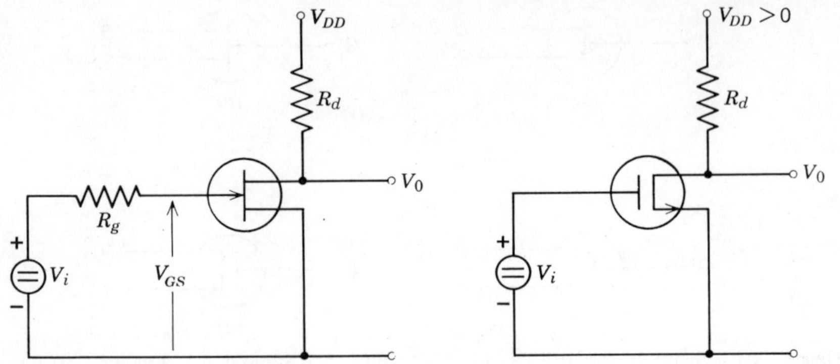

Figure 4.2. (*a*) JFET amplifier or logic circuit with gate protection. (*b*) MOSFET amplifier or logic circuit.

$V_0 > V_{GS} - V_p$, the device is in pinch-off, and the transfer characteristic is

$$V_0 = V_{DD} - K(V_{GS} - V_p)^2 R_d \tag{4.7}$$

This part of the characteristic ends if the device is lifted out of pinch-off. The borderline lies at $V_0 = V_{GS} - V_p$. The equation for V_0 is

$$V_0 = V_{DD} - KV_0^2 R_d \qquad \text{or} \qquad V_0^2 + \frac{V_0}{KR_d} - \frac{V_{DD}}{KR_d} = 0 \tag{4.7a}$$

which has the solution

$$V_0 = V_{GS} - V_p = -\frac{1}{2KR_d} + \left[\frac{1}{(2KR_d)^2} + \frac{V_{DD}}{KR_d}\right]^{1/2} \tag{4.8}$$

(we use here the plus sign, since V_0 must be positive).

Example 1. If $V_p = -4.00$ V, $V_{DD} = 20$ V, $K = 1$ mA/V^2, and $R_d = 5000\ \Omega$. Find the values of V_0 and V_{GS} when the device is just in pinch-off.

Answer.

$$V_0 = V_{GS} - V_T = -0.10 + (0.01 + 4.00)^{1/2} = 1.90 \text{ V}$$

$$V_{GS} = -4.00 + 1.90 = -2.10 \text{ V}$$

Now we look at the case in which the device is not in pinch-off. Then the transfer characteristic follows from the equation

$$I_D = K\left[2(V_{GS} - V_p)V_0 - V_0^2\right] = \frac{V_{DD} - V_0}{R_d} \tag{4.9}$$

leading to the equation

$$V_0^2 - 2\left(V_{GS} - V_p + \frac{1}{2KR_d}\right)V_0 + \frac{V_{DD}}{KR_d} = 0 \tag{4.9a}$$

which has the solution*

$$V_0 = \left(V_{GS} - V_p + \frac{1}{2KR_d}\right) - \left[\left(V_{GS} - V_p + \frac{1}{2KR_d}\right)^2 - \frac{V_{DD}}{KR_d}\right]^{1/2} \tag{4.10}$$

Often

$$\left(V_{GS} - V_p + \frac{1}{2KR_d}\right)^2 \gg \frac{V_{DD}}{KR_d}.$$

Putting $a = V_{GS} - V_p + 1/2KR_d$ and $b = V_{DD}/KR_d$, then $a^2 \gg b$, and

$$a - (a^2 - b)^{1/2} = a - a\left(1 - \frac{b}{a^2}\right)^{1/2} \cong a - a\left(1 - \frac{b}{2a^2}\right) = \frac{b}{2a}$$

so

$$V_0 \cong \frac{V_{DD}/KR_d}{2\left(V_{GS} - V_p + \dfrac{1}{2KR_d}\right)} = \frac{V_{DD}}{2K(V_{GS} - V_p)R_d + 1} = \frac{V_{DD}R_{\text{on}}}{R_d + R_{\text{on}}} \tag{4.10a}$$

where

$$R_{\text{on}} = \frac{1}{2K(V_{GS} - V_p)} \tag{4.10b}$$

Often $R_d \gg R_{\text{on}}$, and then V_0 can be further approximated as

$$V_0 \cong \frac{V_{DD}R_{\text{on}}}{R_d} \tag{4.10c}$$

*Here we need the minus sign, for otherwise $V_0 > V_{GS} - V_p$ and the device is not in the non-pinch-off condition.

We now illustrate the relative merits of these approximation methods with some examples.

Example 2. $V_{DD}=20$ V, $K=1.0$ mA/V^2, $R_d=5000\ \Omega$, $V_{GS}=0.70$ V, and $V_p=-3.00$ V. Find V_0.

Answer. Equation (4.10) yields $V_0=0.57$ V, Eq. (4.10a) yields $V_0=0.53$ V, and Eq. (4.10c) yields $V_0=0.54$ V, since

$$R_{\text{on}}=\frac{1}{2K(V_{GS}-V_p)}=\frac{1}{2\times 10^{-3}\times 3.70}=135\ \Omega$$

Example 3. $V_{DD}=20$ V, $K=1.0$ mA/V^2, $R_d=3000\ \Omega$, $V_{GS}=0.70$ V, and $V_p=-3.00$ V. Find V_0.

Answer. Equation (4.10) yields $V_0=0.99$ V, Eq. (4.10a) yields $V_0=0.86$ V, and Eq. (4.10c) yields $V_0=0.90$ V.

Example 4. $V_{DD}=20$ V, $K=1.0$ mA/V^2, $R_d=2000\ \Omega$, $V_{GS}=0.70$ V, and $V_p=-3.00$ V. Find V_0.

Answer. Equation (4.10) yields $V_0=1.58$ V, Eq. (4.10a) yields $V_0=1.26$ V, and Eq. (4.10c) yields $V_0=1.35$ V.

Example 5. $V_{DD}=20$ V, $K=1.0$ mA/V^2, $V_{GS}=V_{GS0}=0.70$ V, and $V_p=-3.00$ V. Find the value of R_d for which the device is just saturated.

Answer.

$$R_d=\frac{V_{DD}-V_0}{K(V_{GS}-V_p)^2}=\frac{V_{DD}-(V_{GS}-V_p)}{K(V_{GS}-V_p)^2}=1190\ \Omega.$$

It is clear from all these examples that sometimes the full Eq. (4.10) must be used and that Eq. (4.10c) is slightly more accurate than Eq. (4.10a).

If one wants the device to operate *deep* in the ohmic regime at $V_{GS}=V_{GS0}=0.70$ V, one must require

$$I_D=\frac{V_{DD}-V_0}{R_d}\cong\frac{V_{DD}}{R_d}\ll K(V_{GS0}-V_p)^2 \tag{4.11}$$

or

$$R_d \gg \frac{V_{DD}}{K(V_{GS0} - V_p)^2} \tag{4.11a}$$

Under these conditions the device acts as a resistance R_{on}, and V_0 is given approximately by Eq. (4.10c).

4.2b *p*-Channel JFET

This case is the same as the one considered in Section 4.2a, except that all polarities are the opposite, most > signs become < signs and vice versa, and $V_p > 0$. The theory discussed in Section 4.2a must now be transformed.

The device is off if $V_i > V_p$. The device is on if a sufficiently large negative voltage V_i is applied which drives the device into gate current. In this case $V_{GS} = V_{GS0} \cong -0.70$ V, and the condition for V_i is $V_i \ll V_{GS}$. To make the device operate in the ohmic mode, R_d must be chosen so that an appropriately modified form of Eq. (4.11a) is satisfied. The equation for R_{on} is, since $V_{GS} - V_p < 0$,

$$R_{\text{on}} = \frac{1}{g_{d0}} = -\frac{1}{2K(V_{GS} - V_p)} \tag{4.12}$$

If the device is not in pinch-off, Eq. (4.10) must be replaced by*

$$V_0 = \left(V_{GS} - V_p - \frac{1}{2KR_d}\right) + \left[\left(V_{GS} - V_p - \frac{1}{2KR_d}\right)^2 + \frac{V_{DD}}{KR_D}\right]^{1/2} \tag{4.13}$$

since $V_{GS0} - V_p$ and V_{DD} are negative because Eq. (4.9) becomes

$$I_D = -K\left[2(V_{GS} - V_p)V_0 - V_0^2\right] = \frac{V_{DD} - V_0}{R_d}$$

Equation (4.10a) becomes

$$V_0 = V_{DD}\frac{R_{\text{on}}}{R_d + R_{\text{on}}} \tag{4.13a}$$

and, for $R_{\text{on}} \ll R_d$, we have instead of Eq. (4.10c),

$$V_0 = V_{DD}\frac{R_{\text{on}}}{R_d} \tag{4.13b}$$

*Here we need the plus sign, for otherwise $V_0 < V_{GS} - V_p$ and the device is not in the non-pinch-off condition.

4.2c *n*-Channel MOSFET with $V_T < 0$

This case is quite similar to the one Section 4.2a, with two exceptions:

1. There is no restriction on V_i, so that it is always possible to drive the device into the ohmic mode of operation by making V_i sufficiently large.

2. There is no gate current, so no protective resistor R_g is needed. This leads to Fig. 4.2*b*.

One should not make R_d too small, since the drain current in the ohmic regime

$$I_D = \frac{V_{DD} - V_0}{R_d} \cong \frac{V_{DD}}{R_d} \tag{4.14}$$

would become too large. We require for ohmic operation that

$$K(V_i - V_T)^2 \gg \frac{V_{DD}}{R_d} \tag{4.15}$$

from which the value of V_i for a given R_d follows. Except for this, the discussion is the same as for the case considered in Section 4.2a.

4.2d *n*-Channel MOSFET with $V_T > 0$

This case is quite similar to the one discussed in Section 4.2c; the only difference is that $V_T > 0$. The device is off for $V_i < V_T$ and on for $V_i > V_T$. The conditions for operation in the ohmic regime are again given by Eqs. (4.14) and (4.15).

4.2e *p*-Channel MOSFET

Since the *p*-channel MOSFET has $V_T < 0$, and all polarities are opposite those discussed in Section 4.2d, the device is off for $V_i > V_T$ and on for $V_i < V_T$. The conditions for operating in the ohmic regime are again given by Eq. (4.14), but Eq. (4.15) must be replaced by

$$-K(V_i - V_T)^2 \ll \frac{V_{DD}}{R_d} \tag{4.16}$$

because the current is negative and V_{DD} is negative.

4.2f Cascaded Logic FET Circuits

Often one wants to use an FET logic circuit to drive a similar logic circuit. This means that the following requirements must be met: If one stage is turned off, the next stage must be driven into the ohmic regime, and if one stage is driven into the ohmic regime, the next stage must be turned off. All the circuits in Sections 4.2a through 4.2e meet the first requirement, but only the circuits in Sections 4.2d through 4.2e meet the second. Therefore only enhancement MOSFETs can be used in logic chains. For this reason most MOSFET integrated logic circuits use enhancement-mode MOSFETs.

We illustrate this with an example. The circuits in Section 4.2d have $V_T > 0$. It should be possible to design the preceding similar circuit in such a way that, if it is driven into the ohmic regime, then $V_0 < V_T$. The following circuit would then be turned off. This is not possible for the circuits in Section 4.2c, since $V_T < 0$ and $V_0 > 0$.

4.3 INTEGRATED CIRCUIT INVOLVING TWO ENHANCEMENT-MODE MOSFETS

Let two MOSFETs with a p-type channel have the same value of V_T but different values of K, which is achieved by making the two devices on the same silicon chip with widely different w/L ratios; here w is the width and L is the length of the channel. One of the MOSFETs is used as an amplifier, and the other one is used as a load resistor by connecting its gate to the drain (Fig. 4.3).

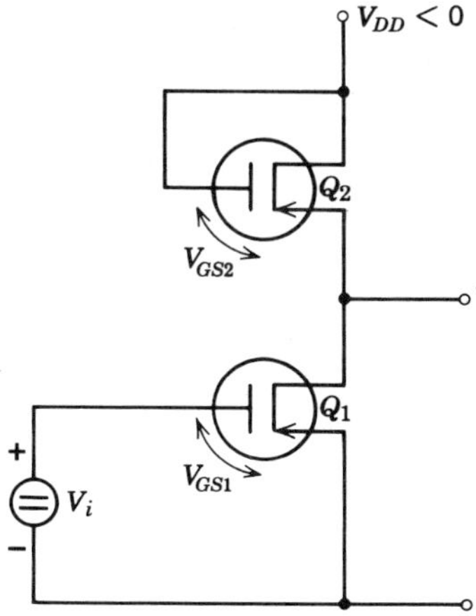

Figure 4.3. Enhancement-mode p-channel MOSFET with enhancement-mode load.

The advantages of this circuit arrangement are:

1. No resistors are needed, which simplifies integrated manufacturing.

2. The circuit works as a linear amplifier for a rather wide range of operating conditions.

3. The circuit works extremely well in logic circuits.

We now discuss the circuit in greater detail.

4.3a The Circuit as an Amplifier

Since the two devices are in series, they have the same current. Hence, for two p-channel devices,

$$I_D = -K_1(V_{GS1} - V_T)^2 = I_{D2} = -K_2(V_{GS2} - V_T)^2 \tag{4.17}$$

or

$$(V_{GS1} - V_T)\left(\frac{K_1}{K_2}\right)^{1/2} = V_{GS2} - V_T \tag{4.17a}$$

as long as both devices operate at pinch-off.

Since the gate of Q_2 is connected to the drain, and Q_2 is an enhancement-mode device, it always operates at pinch-off. However, Q_1 operates at pinch-off for a certain range of values of V_i, which will be determined in a moment.

We must bear in mind that

$$V_{GS1} = V_i \qquad V_{GS2} = V_{DD} - V_0 \tag{4.18}$$

Substitution into Eq. (4.17a) yields

$$(V_i - V_T)\left(\frac{K_1}{K_2}\right)^{1/2} = V_{DD} - V_0 - V_T \qquad V_0 = V_{DD} - V_T - (V_i - V_T)\left(\frac{K_1}{K_2}\right)^{1/2} \tag{4.19}$$

We thus see that V_0 varies linearly with V_i, so the circuit acts as a *linear* amplifier when Q_1 operates under pinch-off conditions and is not turned off. We also see that the circuit has a voltage gain

$$g_v = -\left(\frac{K_1}{K_2}\right)^{1/2} = -\left(\frac{w_1}{L_1}\frac{L_2}{w_2}\right)^{1/2} \tag{4.19a}$$

which can be made quite large by proper choice of the geometries; here w_1 and w_2 are the channel widths, and L_1 and L_2 the channel lengths. Finally the output voltage varies as $-V_i$, indicating a 180° phase difference.

For what range of V_i is the circuit linear? The limits of linear operation are:

1. If $V_i = V_T$, Q_1 is just off. Substitution into Eq. (4.19) yields

$$V_0 = V_{DD} - V_T \tag{4.20}$$

2. If $V_0 = V_i - V_T$, Q_1 is operating just at pinch-off. Substitution into Eq. (4.19) gives

$$V_i - V_T = V_{DD} - V_T - \left(\frac{K_1}{K_2}\right)^{1/2}(V_i - V_T)$$

or

$$V_i - V_T = \frac{V_{DD} - V_T}{1 + (K_1/K_2)^{1/2}} = V_0 \tag{4.21}$$

The device is thus linear for

$$V_T + \frac{V_{DD} - V_T}{1 + (K_1/K_2)^{1/2}} \leqslant V_i \leqslant V_T \tag{4.22}$$

The most suitable dc operating point for Q_1 is

$$V_{GS1} = V_T + \frac{1}{2}\,\frac{V_{DD} - V_T}{1 + (K_1/K_2)^{1/2}} \tag{4.22a}$$

and the largest possible ac amplitude v_i is

$$v_i = -\frac{1}{2}\,\frac{V_{DD} - V_T}{1 + (K_1/K_2)^{1/2}} \tag{4.22b}$$

(Note that $V_{DD} - V_T$ is negative, whereas v_i should be a positive number.) The corresponding range for V_0 is

$$\frac{V_{DD} - V_T}{1 + (K_1/K_2)^{1/2}} \geqslant V_0 \geqslant V_{DD} - V_T \tag{4.23}$$

and the output amplitude is

$$v_0 = \frac{1}{2}(V_{DD} - V_T)\frac{(K_1/K_2)^{1/2}}{1+(K_1/K_2)^{1/2}} \tag{4.23a}$$

indicating that v_0 is 180° out of phase with respect to v_i.

Example 1. If $K_1/K_2 = 25$, $V_{DD} = -27$ V, and $V_T = -3$ V find the dc gate bias V_{GS1} for maximum ac gate amplitude, the ac gate amplitude v_i, and the ac output amplitude v_0.

Answer.

$$V_{GS1} = -3.0 + \frac{1}{2}\frac{(-27+3)}{6} = -3.0 - 2.0 = -5.0 \text{ V}$$

$$v_i = -\frac{1}{2}\frac{(-27+3)}{6} = 2.0 \text{ V}$$

$$v_0 = +\frac{1}{2}(-27+3)\frac{5}{6} = -10.0 \text{ V}$$

so $v_0/v_i = -5$ as expected. The device is linear for $-7.0 \text{ V} < V_{GS1} < -3.0$ V.

While the circuit has interesting features as an amplifier, it has even more interesting features as a logic circuit. We discuss them in Section 4.3b.

4.3b The Circuit as a Logic Circuit

Since all voltages in the circuit are negative, and logic thus far has been mostly defined for positive voltages (*positive* logic), we have to repeat the definition of logic for negative voltages (*negative* logic). A 0 is now defined as a negative voltage close to 0 V, whereas a 1 is defined as a voltage that is strongly negative; say, $V > -2$ V is a 0 and $V < -10$ V is a 1.

We now see that in our circuit a 0 input gives a 1 output and a 1 input gives a 0 output, so the circuit acts as an *inverter*. Let the devices have the same turn-on voltage $V_T < 0$; then $V_i > V_T$ is called a 0 input, and $V_i \ll V_T$ is called a 1 input.

Now when the circuit has a 0 input, the output voltage is a 1. For $V_i > V_T$, according to Eq. (4.20),

$$V_0 = V_{DD} - V_T \tag{4.24}$$

and the device is off. When, however,

$$V_i - V_T \ll \frac{V_{DD} - V_T}{1 + (K_1/K_2)^{1/2}} \tag{4.25}$$

(1 input), the device operates in the ohmic region and V_0 is quite close to zero, say $V_0 > V_T$, so the output is a 0 as previously stated.

We now calculate V_0 for this case. Since both devices are p-type and Q_1 operates in the non-pinch-off region, Eq. (4.17) must be replaced by

$$I_{D1} = -K_1\left[2(V_i - V_T)V_0 - V_0^2\right] = I_{D2} = -K_2(V_{DD} - V_T - V_0)^2 \tag{4.26}$$

This equation can be written as

$$V_0^2 - 2\frac{K_1(V_i - V_T) + K_2(V_{DD} - V_T)}{K_1 + K_2}V_0 + \frac{K_2(V_{DD} - V_T)^2}{K_1 + K_2} = 0 \tag{4.26a}$$

which has the solution*

$$V_0 = \frac{K_1(V_i - V_T) + K_2(V_{DD} - V_T)}{K_1 + K_2} + \left\{\left[\frac{K_1(V_i - V_T) + K_2(V_{DD} - V_T)}{K_1 + K_2}\right]^2 - \frac{K_2(V_{DD} - V_T)^2}{K_1 + K_2}\right\}^{1/2} \tag{4.27}$$

Often

$$\left[\frac{K_1(V_i - V_T) + K_2(V_{DD} - V_T)}{K_1 + K_2}\right]^2 \gg \frac{K_2(V_{DD} - V_T)^2}{K_1 + K_2}$$

We put

$$a = \frac{K_1(V_i - V_T) + K_2(V_{DD} - V_T)}{K_1 + K_2} \quad \text{and} \quad b = \frac{K_2(V_{DD} - V_T)^2}{K_1 + K_2}$$

*Here we need the plus sign, for otherwise $V_0 < V_i - V_T$ and Q_1 is not in the non-pinch-off condition.

then, for $a^2 \gg b$ and $a<0$,

$$a+(a^2-b)^{1/2}=a-a\left(1-\frac{b}{a^2}\right)^{1/2} \cong a-a\left(1-\frac{b}{2a^2}\right)=\frac{b}{2a}$$

so

$$V_0 \cong \frac{K_2(V_{DD}-V_T)^2}{2K_1(V_i-V_T)+2K_2(V_{DD}-V_T)}=I_{D0}R_{\text{on}}\frac{R_d}{R_d+R_{\text{on}}} \tag{4.27a}$$

This has the following simple interpretation. In the ohmic regime Q_1 has a resistance

$$R_{\text{on}}=-\frac{1}{2K_1(V_i-V_T)}>0$$

When the output of Q_1 is short-circuited, Q_2 produces a current

$$I_{D0}=-K_2(V_{DD}-V_T)^2<0$$

and the device Q_2 has an internal resistance

$$R_d=-\frac{1}{2K_2(V_{DD}-V_T)}>0$$

so V_0 is due to the flow of a current I_{D0} through the parallel connection between R_d and R_{on}.

Often $R_{\text{on}} \ll R_d$ and in this case Eq. (4.27a) can be approximated as

$$V_0=I_{D0}R_{\text{on}}=\frac{K_2(V_{DD}-V_T)^2}{2K_1(V_i-V_T)} \tag{4.27b}$$

so V_0 comes about because the current I_{D0} flows into R_{on}.

Example 2. If $V_T=-3.0$ V, $V_{DD}=-21.0$ V, $K_1=1.25$ mA/V^2, and $K_2=0.05$ mA/V^2, find the output voltage for

(a) $V_i=-18.0$ V
(b) $V_i=-10.0$ V
(c) $V_i=-7.0$ V
(d) $V_i=-6.0$ V
(e) $V_i>-3.0$ V

Compare Eqs. (4.27), (4.27a), and (4.27b) for (a) through (d).

Answer.

(a) Equation (4.27) yields $V_0 = -0.42$ V, Eq. (4.27a) yields $V_0 = -0.41$ V, and Eq. (4.27b) yields $V_0 = -0.43$ V.
(b) Equation (4.27) yields $V_0 = -0.89$ V, Eq. (4.27a) yields $V_0 = -0.84$ V, and Eq. (4.27b) yields $V_0 = -0.93$ V.
(c) Equation (4.27) yields $V_0 = -1.69$ V, Eq. (4.27a) yields $V_0 = -1.37$ V, and Eq. (4.27b) yields $V_0 = -1.62$ V.
(d) Equation (4.27) yields $V_0 = -3.00$ V, Eq. (4.27a) yields $V_0 = -1.74$ V, and Eq. (4.27b) yields $V_0 = -2.16$ V. The device is now just in pinch-off.
(e) For $V_i > V_T$, $V_0 = V_{DD} - V_T = -21 + 3 = -18$ V and the device is off.

The conclusion is that the approximations (4.27a) and (4.27b) can be used only if V_0 is relatively small; otherwise Eq. (4.27) should be used. We also note that Eq. (4.27b) is somewhat more accurate than Eq. (4.27a).

If both devices have n-type channels, the theory remains the same, but the plus sign in front of the square-root term in Eq. (4.27) becomes a minus sign,* and hence

$$V_0 = \frac{K_1(V_i - V_T) + K_2(V_{DD} - V_T)}{K_1 + K_2} - \left\{ \left[\frac{K_1(V_i - V_T) + K_2(V_{DD} - V_T)}{K_1 + K_2} \right]^2 - \frac{K_2(V_{DD} - V_T)^2}{K_1 + K_2} \right\}^{1/2} \tag{4.27c}$$

4.4 PULSE RESPONSE OF FETS

An exact calculation of the pulse response of FETs is rather difficult, since one must not only take into account the response of the input and output circuits, but also the feedback effect caused by the drain-gate capacitance C_{dg} and the response of the FET itself.

In Sections 4.4a and b we discuss the FET pulse response and show that the turn-on response is much faster than the turn-off response. In Section 4.4c we discuss a circuit that has fast turn-on *and* turn-off response.

4.4a Pulse Response of MOSFETs

To illustrate the problem, we turn to Fig. 4.4*a*, which shows an enhancement-mode n-channel MOSFET ($V_T > 0$) driven by a pulse signal $V(t)$, with a resistance R_g in the gate lead, a resistance R_d in the drain lead, and

*Otherwise $V_0 > V_i - V_T$ and Q_1 is not in the non-pinch-off condition.

a load capacitance C_d which is mostly due to the input capacitances of subsequent FET stages. The equivalent circuit is shown in Fig. 4.4*b*, which also shows the device capacitances C_{gs} and C_{dg}; the capacitance C_{ds} has been absorbed into the load capacitance C_d.

We now assume for simplicity that R_d is taken quite small, so the feedback via the capacitance C_{dg} can be neglected. We then have for the response of the input circuit, if $V(t)=0$ for $t<0$ and $V(t)=V_1$ for $t>0$,

$$V_g(t)=V_1\left[1-\exp\left(-\frac{t}{\tau_g}\right)\right] \tag{4.28}$$

where

$$\tau_g=C_{\text{in}}R_g=R_g(C_{gs}+C_{dg}) \tag{4.28a}$$

and $C_{\text{in}}=C_{gs}+C_{dg}$ is the input capacitance. This is the turn-on response of the MOSFET.

Next we consider another simple case in which $C_d=0$ and R_d is somewhat larger. The input capacitance is then larger, because of the Miller effect, and is given by

$$C_{\text{in}}=C_{gs}+C_{dg}(1+g_mR_d) \tag{4.28b}$$

where g_m is the transconductance. Equation (4.28) is still valid, but the second half of Eq. (4.28a) must be replaced by

$$\tau_g=R_g[C_{gs}+C_{dg}(1+g_mR_d)] \tag{4.28c}$$

The *intrinsic* response of the FET, that is, the time it takes to form a

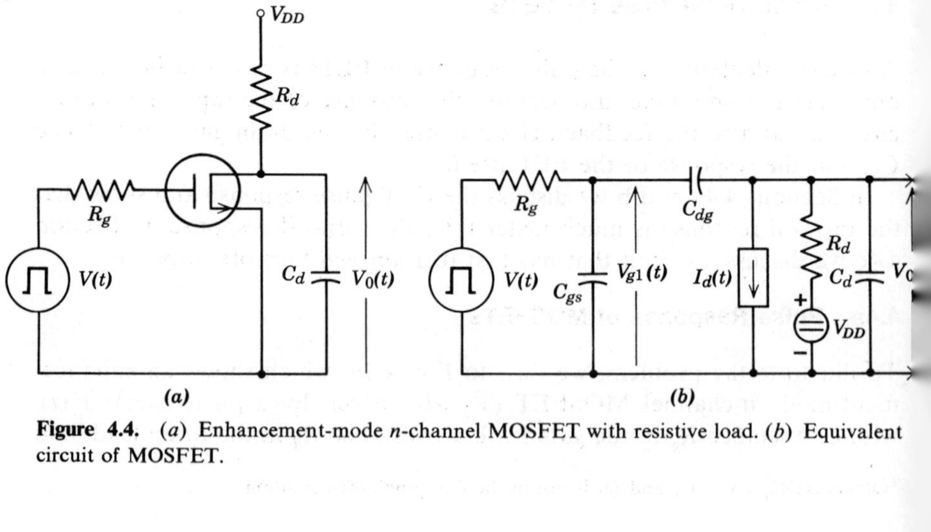

Figure 4.4. (*a*) Enhancement-mode *n*-channel MOSFET with resistive load. (*b*) Equivalent circuit of MOSFET.

conducting channel between the source and the drain after the gate voltage has been turned on, is usually so short ($<10^{-9}$ s) that it can be neglected. We thus assume that $I_d(t)$ follows $V_g(t)$ instantaneously, so

$$I_d(t) = K[V_g(t) - V_T]^2 \tag{4.29}$$

This current generator operates on the output circuit. If the time constant $\tau_d = C_d R_d$ is much smaller than the time constant τ_g, we have in good approximation

$$V_0(t) = V_{DD} - I_d(t) R_d = V_{DD} - K R_d [V_g(t) - V_T]^2 \tag{4.29a}$$

so the input pulse causes a decrease in the output voltage, as expected. Here $V_g(t)$ is given by Eq. (4.28), and τ_g by Eq. (4.28c).

Next we add a small refinement. We see from the circuit in Fig. 4.4*b* that the capacitance C_{dg} gives rise to a feedthrough effect. As long as the device is off, this feedthrough effect gives rise to a positive spike in the output caused by $V_g(t)$. After the device has been turned on, the current generator $I_d(t)$ takes over and drives $V_0(t)$ lower, ending up as in Eq. (4.29a).

4.4b Charging and Discharging of the Load Capacitance C_d.

We next discuss another simple case, in which R_g is quite small and R_d much larger, so

$$\tau_g = R_g(C_{gs} + C_{dg}) \ll C_d R_d = \tau_d$$

The transient then consists mainly of the charging and discharging of the load capacitance C_d.

We first consider the turn-on transient of the circuit by assuming that $V(t) = 0$ for $t < 0$ and $V(t) = V_1$ for $t > 0$, where V_1 is so large that the device is driven deeply into the ohmic regime. Since τ_g is so small, we assume that $V_g(t) = V(t)$.

We represent the (I_D, V_{DS}) characteristic of the device by a linear model consisting of a pinch-off regime for $V_0 \geqslant \frac{1}{2}(V_1 - V_T) = V_{p0}$, where $I_D = I_{D0}$ = constant, and an ohmic regime, for $V_0 \leqslant \frac{1}{2}(V_1 - V_T)$, for which $V_0 = I_D R_{on}$, where $1/R_{on} = 2K(V_1 - V_T)$ (Fig. 4.5*a*).

First the device is pinched off, and the equivalent circuit is as shown in Fig. 4.5*b*. For $t > 0$ the differential equation of the circuit is

$$C_d \frac{dV_0}{dt} = \frac{V_{DD} - I_{D0} R_d - V_0}{R_d} \tag{4.30}$$

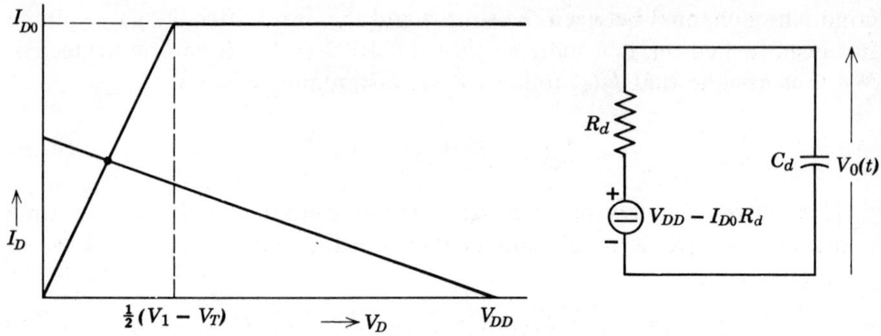

Figure 4.5. (*a*) Schematic (I_D, V_D) characteristic with load line, showing that the device is operating in the ohmic regime. (*b*) Equivalent circuit when the device is operating at constant current (pinch-off mode).

or, by dividing by C_d and putting $\tau_d = C_d R_d$,

$$\frac{dV_0}{dt} + \frac{V_0}{\tau_d} = \frac{V_{DD} - I_{D0}R_d}{\tau_d} \tag{4.30a}$$

These equations hold as long as the device is pinched off, or $V_0 \geqslant V_{p0} = \frac{1}{2}(V_1 - V_T)$. The initial condition is $V_0(t) = V_{DD}$ at $t=0$. Here $I_{D0}R_d$ is usually large in comparison with V_{DD}.

The solution of this equation is

$$V_0(t) = V_{DD} - I_{D0}R_d + A \exp\left(-\frac{t}{\tau_d}\right) \tag{4.30b}$$

since $V_0(t) = V(\infty) = V_{DD} - I_{D0}R_d$ is a solution to Eq. (4.30a). Substituting the initial condition at $t=0$ yields $A = I_{D0}R_d$, or

$$V_0(t) = V_{DD} - I_{D0}R_d\left[1 - \exp\left(-\frac{t}{\tau_d}\right)\right] \tag{4.30c}$$

This part of the solution terminates when $V_0 = V_{p0}$. Let this be at $t = t_1$. Substituting $V_0(t) = V_{p0}$ at $t = t_1$,

$$V_{p0} = V_{DD} - I_{D0}R_d + I_{D0}R_d \exp\left(-\frac{t_1}{\tau_d}\right) \tag{4.31}$$

from which it follows that

$$t = t_1 = \tau_d \ln\left(\frac{I_{D0}R_d}{I_{D0}R_d - V_{DD} + V_{p0}} \right) \tag{4.31a}$$

We bear in mind that $I_{D0}R_d \gg V_{DD} - V_{p0}$. We can then rewrite Eq. (4.31a) as

$$t_1 = -\tau_d \ln\left(1 - \frac{V_{DD} - V_{p0}}{I_{D0}R_d}\right) \cong \tau_d \frac{V_{DD} - V_{p0}}{I_{D0}R_d} = \frac{C_d(V_{DD} - V_{p0})}{I_{D0}} \tag{4.31b}$$

since $\ln(1-x) \cong -x$ for small x, and $\tau_d = C_d R_d$.

This has a very simple explanation: $C_d(V_{DD} - V_{p0})$ is the charge lost by C_d during the time t_1, and the capacitor is discharged by a constant current I_{D0}; hence t_1 is simply the ratio of the lost charge to the discharging current.

We now investigate the response in the ohmic regime. If we replace the emf V_{DD} by the current generator V_{DD}/R_d and assume $R_d \gg R_{\text{on}}$, we obtain the equivalent circuit in Fig. 4.6.

The differential equation of the circuit is

$$C_d \frac{dV_0}{dt} + \frac{V_0}{R_{\text{on}}} = \frac{V_{DD}}{R_d} \tag{4.32}$$

or, if $\tau_{\text{on}} = C_d R_{\text{on}}$,

$$\frac{dV_0}{dt} + \frac{V_0}{\tau_{\text{on}}} = \frac{V_{DD}}{\tau_d} \tag{4.32a}$$

with the initial condition $V_0 = V_{p0}$ at $t = t_1$. Since

$$V_0 = V_0(\infty) = V_{\text{on}} = V_{DD}\frac{\tau_{\text{on}}}{\tau_d} = V_{DD}\frac{R_{\text{on}}}{R_d} \tag{4.32b}$$

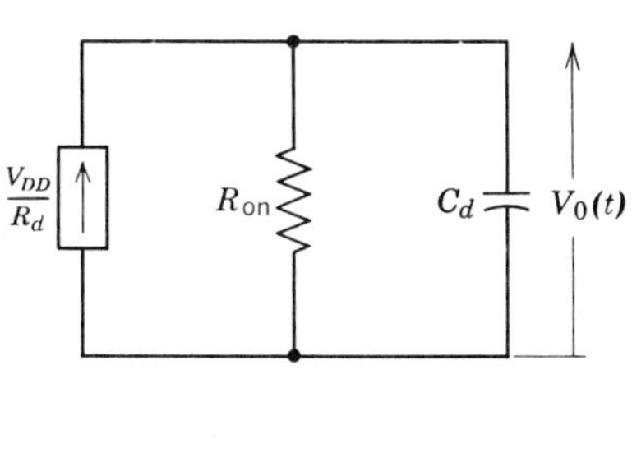

Figure 4.6. Equivalent circuit when the device is operating in the ohmic regime.

is a solution to the equation, and $A\exp[-(t-t_1)/\tau_{\text{on}}]$ is a solution to the homogeneous equation, the solution to Eq. (4.32a) satisfying the initial condition is

$$V_0(t)=V_{\text{on}}\left[1-\exp\left(-\frac{t-t_1}{\tau_{\text{on}}}\right)\right]+V_{p0}\exp\left(-\frac{t-t_1}{\tau_{\text{on}}}\right) \tag{4.32c}$$

We now define the turn-on time as the total time t_2 required for $V_0(t)$ to drop over 90% of its range $V_{DD}-V_{\text{on}}$. Let this occur at $t=t_2$; then $V_0(t_2)=V_{DD}-0.90(V_{DD}-V_{\text{on}})$. Substituting into Eq. (4.32c) yields

$$0.10(V_{DD}-V_{\text{on}})=(V_{p0}-V_{\text{on}})\exp\left(-\frac{t_2-t_1}{\tau_{\text{on}}}\right) \tag{4.32d}$$

or

$$t_2-t_1=\tau_{\text{on}}\ln\left[\frac{V_{p0}-V_{\text{on}}}{0.10(V_{DD}-V_{\text{on}})}\right] \tag{4.32e}$$

We now illustrate this with an example.

Example 1. Let $V_{DD}=20$ V, $V_1=12$ V, $V_T=2$V, $I_{D0}=20$ mA, $R_d=5000$ Ω, and $C_d=20$ pF. Find the response time t_2 for the turn-on transient.

Answer. $V_{p0}=\frac{1}{2}(V_1-V_T)=5$ V, $I_{D0}R_d=100$ V, $\tau_d=C_dR_d=20\times10^{-12}\times5\times10^3=100\times10^{-9}$ s. Then, from Eq. (4.31a),

$$t_1=10^{-7}\ln\frac{100}{85}=16.3\times10^{-9}\text{ s.}$$

If we had used Eq. (4.31b) we would have obtained

$$t_1=\frac{20\times10^{-12}\times15}{20\times10^{-3}}=15\times10^{-9}\text{ s}$$

which is quite close.

Next we evaluate t_2-t_1 as follows. $I_{D0}=K(V_1-V_T)^2$, $20\times10^{-3}=100\,K$, or $K = 0.2\ \mathrm{mA/V^2}$.

$$R_{\text{on}} = \frac{1}{2K(V_1 - V_T)} = \frac{1}{4\times 10^{-3}} = 250\ \Omega$$

$$V_{\text{on}} = V_{DD}\frac{R_{\text{on}}}{R_d} = 20\times\frac{250}{5000} = 1.0\ \text{V}$$

$$\tau_{\text{on}} = C_d R_{\text{on}} = 20\times 10^{-12}\times 250 = 5\times 10^{-9}\ \text{s}$$

$$t_2 - t_1 = 5\times 10^{-9}\ln\left(\frac{4}{1.9}\right) = 3.7\times 10^{-9}\ \text{s}$$

$$t_2 = 16.3\times 10^{-9} + 3.7\times 10^{-9} = 20.0\times 10^{-9}\ \text{s}$$

The actual response time is slightly longer, for the piecewise linear characteristic in Fig. 4.5*a* is higher than the actual characteristic for the full voltage range. This difference is not very great, however; the value of t_2 obtained here is probably accurate within 10%.

We now discuss the turn-off response. Since $V(t) = V_1$ for $t < 0$ and $V(t) = 0$ for $t > 0$, we have $V_0(0^+) = V_{\text{on}}$ and $I(t) = 0$ for $t > 0^+$. Hence the output equivalent circuit is as shown in Fig. 4.7, and the output response is governed by the equation

$$\frac{dV_0}{dt} + \frac{V_0}{\tau_d} = \frac{V_{DD}}{\tau_d} \tag{4.33}$$

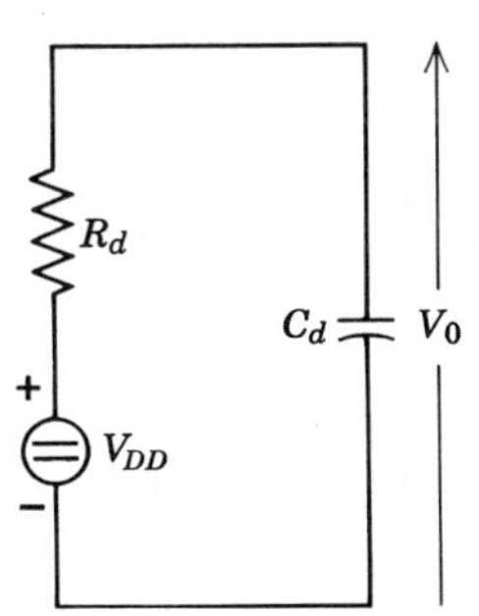

Figure 4.7. Equivalent circuit when the device is off.

with $V_0(0^+) = V_{on}$. The solution is

$$V_0(t) = V_{on} \exp\left(-\frac{t}{\tau_d}\right) + V_{DD}\left[1 - \exp\left(-\frac{t}{\tau_d}\right)\right]$$

$$= V_{DD} - (V_{DD} - V_{on}) \exp\left(-\frac{t}{\tau_d}\right) \tag{4.33a}$$

since $V_0(t) = V_0(\infty) = V_{DD}$ is a solution to Eq. (4.33).

We now investigate how long it takes $V_0(t)$ to change from V_{on} to $V_{on} + 0.90(V_{DD} - V_{on})$. Let this occur at $t = t_3$; then

$$V_{on} + 0.90(V_{DD} - V_{on}) = V_{DD} - (V_{DD} - V_{on}) \exp\left(-\frac{t_3}{\tau_d}\right) \tag{4.33b}$$

or

$$\exp\left(-\frac{t_3}{\tau_d}\right) = 0.10 \qquad t_3 = \tau_d \ln 10 = 2.3\tau_d \tag{4.34}$$

This is much larger than the turn-on time t_2, as seen from the following example.

Example 2. Find the turn-off time t_3 for the circuit discussed in Example 1.

Answer. Since $\tau_d = 10^{-7}$ s, $t_3 = 2.3\tau_d = 230 \times 10^{-9}$ s which is about 10 times larger than t_2.

We have discussed here the case in which R_g is relatively small so that the transient consists of charging or discharging C_d. When R_g is much larger, the problem becomes so much more complicated that it is beyond the scope of this book. The case involving small C_d and relatively large values of R_g and R_d was considered in Section 4.4a.

4.4c Complementary Symmetry MOSFET Circuits (CMOS)

This circuit is shown in Fig. 4.8; it consists of a *p*-channel and an *n*-channel enhancement-mode MOSFET pair connected in series for direct current, with the gates connected in parallel. C_0 is the load capacitance of the circuit due to subsequent MOS stages. We shall see that the circuit has practically zero current, except during switching.

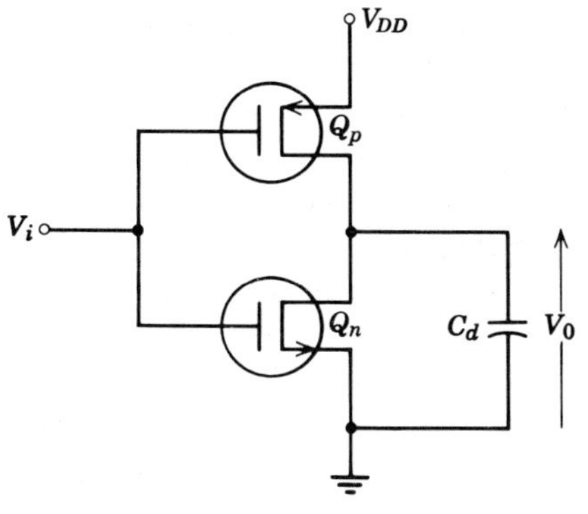

Figure 4.8. CMOS circuit.

We see by inspection that the n-channel device has gate-source and drain-source voltages

$$V_{GSn} = V_i \qquad V_{DSn} = V_0 \tag{4.35}$$

respectively, whereas the p-channel device has voltages

$$V_{GSp} = V_i - V_{DD} \qquad V_{DSp} = V_0 - V_{DD} \tag{4.36}$$

If V_i is at a logical 1, for example, $V_i \cong V_{DD}$, then Q_p, since it is an enhancement-mode device, is turned off (i.e., it operates at a leakage current of about 10^{-12} A), while Q_n is turned on. However, since the current is determined by Q_p, Q_n operates deeply in the ohmic regime. We now have the (I_D, V_0) characteristics shown in Fig. 4.9a. The Q_n characteristic passes through the point $I_D=0$, $V_0=0$ V, and the Q_p characteristic through the point $I_D=0$, $V_0=V_{DD}$. The operating point is the point of

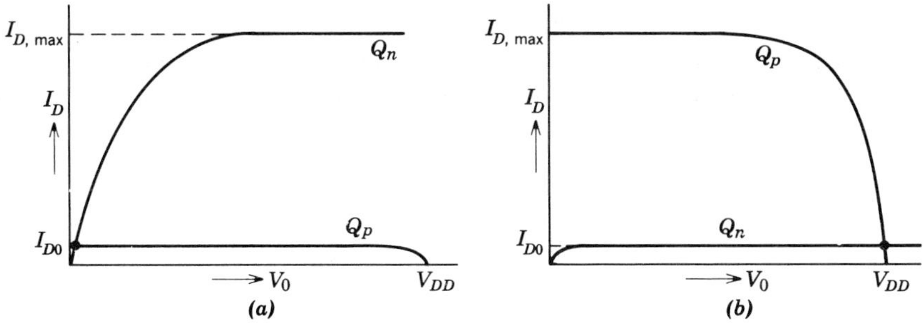

Figure 4.9. (a) (I_D, V_0) characteristics when Q_n is on and Q_p is off. (b) (I_D, V_0) characteristics when Q_p is on and Q_n is off.

intersection of the two characteristics (i.e., V_0 is very close to 0 V), so the output is a logical 0.

If V_i is a logical 0, for example, $V_i \cong 0$, then Q_n is turned off (i.e., operates at a leakage current of the order of 10^{-12} A) and Q_p is turned on. However, since the current is determined by Q_n, Q_p operates deeply in the ohmic regime. We now have the (I_D, V_0) characteristics shown in Fig. 4.9*b*. The Q_n characteristic again passes through the origin, and the Q_p characteristic passes through the point $I_D = 0$, $V_0 = V_{DD}$. The operating point is the point of intersection of the two characteristics (i.e., V_0 is very close to V_{DD}), so the output is a logical 1. If the two devices are completely symmetrical, they have the same current I_D when off, and the same pinch-off current $I_{D,\max}$ when on.

Since a logical 1 input produces a logical 0 output and a logical 0 input produces a logical 1 output, the circuit is an *inverter*.

The CMOS circuit is very well suited for operation at low voltages, hence low powers. To this end one must make the turn-on voltages V_{T1} and V_{T2} of the two devices close to zero; since V_0 is either 0 V or equal to V_{DD}, this is no handicap. For an *n*-channel device this is not difficult to achieve, for a *p*-channel MOSFET ion implantation can be used to bring the turn-on voltage closer to zero. Care should be taken of course that the devices stay in the enhancement mode.

We now turn to the transients developed during switching. Suppose V_i switches from a logical 1 to a logical 0; then Q_n is switched off, and Q_p is switched on. Since V_0 is first at 0 V, Q_p has a strongly negative drain voltage, $V_{DS} \cong -V_{DD}$, so it is pinched off, that is, $I_D \cong I_{D,\max}$. Hence a large current $I_D(t)$ flows, which charges C_d until it reaches the voltage V_{DD}.

Suppose V_i switches from a logical 0 to a logical 1; then Q_p is switched off and Q_n is switched on. Since at first $V_0 \cong V_{DD}$, a large current flows through Q_n, discharging C_d until it reaches zero voltage. If the devices are fully symmetrical, the currents are equal at identical values of V_{DS}, hence the discharge time is equal to the charging time; this overcomes one of the drawbacks of the regular MOSFET circuit.

We now define the discharging time as the time t_2 needed to discharge C_d from V_{DD} to $0.10V_{DD}$. The total charge displaced during t_2 is thus $0.90C_dV_{DD}$, hence t_2 follows from the relation

$$\int_0^{t_2} I_D(t)\,dt = 0.90C_dV_{DD} \tag{4.37}$$

If we make the slightly optimistic assumption that $I_D(t) = I_{D,\max}$ for $0 < t <$

t_2, we have

$$I_{D,\max}t_2 = 0.90C_dV_{DD} \qquad \text{or} \qquad t_2 = \frac{0.90C_dV_{DD}}{I_{D,\max}} \tag{4.37a}$$

Since $I_D(t) < I_{D,\max}$ part of the time, Eq. (4.37a) somewhat underestimates t_2, but not by an important factor.

To obtain a more accurate estimate of t_2, we approximate the characteristic by a piecewise linear one, as in Section 4.4b. Suppose the devices have turn-on voltages V_{Tp} and V_{Tn}, with $V_{Tp} = -V_{Tn}$. Suppose further that $V_0 = V_{DD}$ for $t = 0^+$ and that C_d discharges through an n-type device for $t > 0$. We finally assume that $V_i = V_{DD}$. Then a piecewise linear characteristic can be defined that has

$$I_D = I_{D,\max} = K(V_{DD} - V_{Tn})^2 \tag{4.38}$$

for $V_0 \geqslant V_{p0} = \frac{1}{2}(V_{DD} - V_{Tn})$, and

$$I_D = \frac{V_0}{R_{\text{on}}} \qquad \text{with } R_{\text{on}} = \frac{1}{2K(V_{DD} - V_{Tn})} \tag{4.38a}$$

for $V_0 \leqslant V_{p0}$.

We now consider the transient for two time intervals $0 \leqslant t \leqslant t_1$ and $t_1 \leqslant t \leqslant t_2$. During the first time interval $I_D = I_{D,\max}$, and during the second interval C_d discharges through R_{on}. We then have, in analogy with Eq. (4.31b) (Section 4.4b),

$$t_1 = \frac{C_d(V_{DD} - V_{p0})}{I_{D,\max}} \tag{4.39}$$

since the charge $C_d(V_{DD} - V_{p0})$ is removed by the current $I_{D,\max}$. Furthermore, in analogy with Eq. (4.32e),

$$t_2 - t_1 = \tau_{\text{on}} \ln\left(\frac{V_{p0}}{0.10V_{DD}}\right) \tag{4.40}$$

where $\tau_{\text{on}} = C_dR_{\text{on}}$, since $V_{\text{on}} = 0$ in this case. If we now compare the value of t_2, obtained by adding Eqs. (4.39) and (4.40), with the value obtained from Eq. (4.37a), we can see how good an approximation Eq. (4.37a) gives.

Example 2. Compare Eq. (4.37a) with Eq. (4.39) and Eq. (4.40) for the following situation: $V_{DD} = 10$ V, $I_{D,\max} = 10$ mA, $V_{Tn} = 2.0$ V, $V_{Tp} = -2.0$ V, and $C_d = 20$ pF.

Answer. Equation (4.37a) yields

$$t_2 = \frac{0.90 C_d V_{DD}}{I_{D,\max}} = \frac{0.90 \times 20 \times 10^{-12} \times 10}{10^{-2}} = 1.8 \times 10^{-8}\ \text{s}$$

Equations (4.39) and (4.40) yield

$$t_1 = \frac{C_d (V_{DD} - V_{p0})}{I_{D,\max}} = \frac{20 \times 10^{-12} \times 6.0}{10^{-2}} = 1.2 \times 10^{-8}\ \text{s}$$

since $V_p = \frac{1}{2}(10 - 2) = 4.0$ V.

$$t_2 - t_1 = \tau_{\text{on}} \ln \frac{V_{p0}}{0.10 V_{DD}} = C_d R_{\text{on}} \ln 4$$

$$R_{\text{on}} = \frac{1}{2K(V_{DD} - V_{Tn})} = \frac{V_{DD} - V_{Tn}}{2K(V_{DD} - V_{Tn})^2} = \frac{V_{DD} - V_{Tn}}{2I_{D,\max}}$$

$$= \frac{8.0}{2 \times 10^{-2}} = 400\ \Omega$$

$$\tau_{\text{on}} = C_d R_{\text{on}} = 20 \times 10^{-12} \times 400 = 0.8 \times 10^{-8}\ \text{s}$$

$$t_2 - t_1 = 0.8 \times 10^{-8} \times 1.39 = 1.1 \times 10^{-8}\ \text{s}$$

$$t_2 = t_1 + t_2 - t_1 = 1.2 \times 10^{-8} + 1.1 \times 10^{-8} = 2.3 \times 10^{-8}\ \text{s}$$

This is relatively close to the result obtained with Eq. (4.37a), so the last result should be reasonably accurate.

4.4d Depletion-Mode Devices as Loads

When the output of the circuit in Fig. 4.3 feeds into a capacitive load, the turn-on transient for Q_1 is quite fast, since Q_1 has a large saturated current which discharges the load capacitance C_d rapidly. During the turn-off transient, however, the capacitance C_d is charged by a relatively small current, and as a consequence the turn-off transient is quite slow. Steps must therefore be taken to improve the situation. We shall see that this can be achieved by replacing Q_2 by a depletion-mode load device with its gate directly connected to its source.

Figure 4.10*a* shows Fig. 4.3 redrawn for two *n*-channel enhancement-mode devices with the output feeding into a capacitance C_d. In the circuit in Fig. 4.10*b*, the enhancement-mode load Q_2 is replaced by a depletion-mode load Q_2', and the output feeds into the capacitance C_d. Figure 4.10*c* shows the characteristics and the operating point for the two cases.

Let Q_1 have a current I_{D1} and let

$$I_{D1} = I_{DS1} = K_1 (V_i - V_{T1})^2 \tag{4.41}$$

as long as Q_1 is saturated; here $V_{T1} > 0$. Let Q_2 have a current I_{D2}, where

$$I_{D2} = K_2 (V_{DD} - V_{T1} - V_0)^2 \tag{4.42}$$

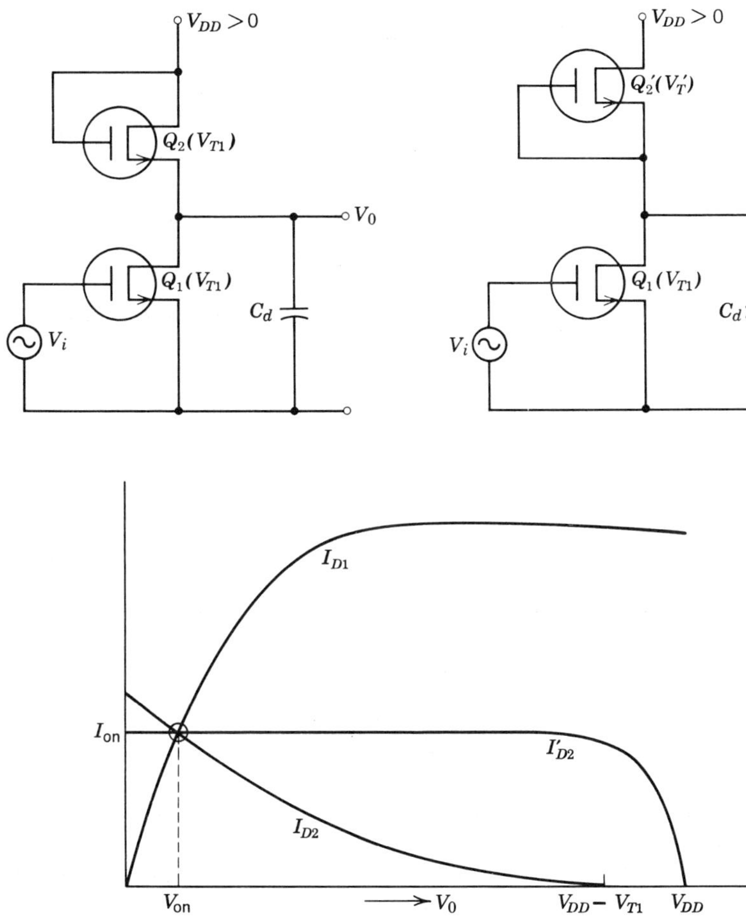

Figure 4.10. (*a*) Enhancement mode *n*-channel MOSFET with *n*-channel enhancement-mode load. (*b*) Enhancement-mode *n*-channel MOSFET with *n*-channel depletion-mode load. (*c*) Characteristics and operating points for the two cases.

and $K_2 \ll K_1$. This device is always saturated, and it is assumed that Q_2 has the same turn-on voltage as Q_1. Finally Q_2' has a current I_{D2}', where

$$I_{D2}' = I_{DS2}' = K_2'(V_T')^2 \tag{4.43}$$

as long as $V_{DD} - V_{on} \geqslant -V_T'$; $V_T' < 0$, because Q_2' is a depletion-mode device.

Since the two devices are in series, their currents are identical in both cases. We now design the circuits such that they have the same operating point (I_{on}, V_{on}); that is,

$$I_{on} = K_2(V_{DD} - V_{T1} - V_{on})^2 = I_{D2}' = K_2'(V_T')^2 \tag{4.44}$$

If Q_1 is turned on, the net discharging current of C_d is $I_{D1} - I_{D2}$ for the circuit in Fig. 4.10*a*, and $I_{D1} - I_{D2}'$ for the circuit in Fig. 4.10*b*. As long as $I_{DS1} \gg I_{on}$, the difference in discharge current is relatively small, so the duration of the two turn-on transients is nearly the same. When Q_1 is turned off, the charging current of C_d is I_{D2} for the circuit in Fig. 4.10*a*, and $I_{D2}' > I_{D2}$ for the circuit in Fig. 4.10*b*. We shall see that the latter results in a considerable speedup of the response.

To evaluate this in detail, we consider first the circuit in Fig. 4.10*b*. If Q_1 is turned off at $t = 0$,

$$C_d \frac{dV_0}{dt} = I_{DS2}' = I_{on} \tag{4.45}$$

with $V_0 = V_{on}$ at $t = 0$, so

$$C_d(V_0 - V_{on}) = I_{on} t \tag{4.45a}$$

Consequently the transient response is 90% complete at $t = t'$, where $V_0 = 0.90(V_{DD} - V_{on}) + V_{on}$; then

$$t_1' = \frac{0.90 C_d (V_{DD} - V_{on})}{I_{on}} \tag{4.45b}$$

We have made the approximation of assuming that $I_{D2}' = I_{DS2}' = I_{on}$ for $V_{on} < V_0 < 0.90 V_{DD}$; at worst this results in only a small error.

We now turn to the circuit in Fig. 4.10*a*. If Q_1 is turned off at $t = 0$, and the transient is 90% complete at $t = t_1$, we have

$$C_d \frac{dV_0}{dt} = I_{D2} = K_2(V_{DD} - V_T - V_0)^2 \tag{4.46}$$

with $V = V_{\text{on}}$ at $t = 0$ and, consequently,

$$\frac{K_2 t_1}{C_d} = \int_{V_{\text{on}}}^{0.90(V_{DD} - V_T - V_{\text{on}}) + V_{\text{on}}} \frac{dV_0}{(V_{DD} - V_T - V_0)^2} = \frac{9}{V_{DD} - V_T - V_{\text{on}}}$$

(4.46a)

since the output changes from V_{on} to $V_{DD} - V_T$. Hence

$$t_1 = \frac{9C_d}{K_2(V_{DD} - V_T - V_{\text{on}})} = \frac{9C_d(V_{DD} - V_T - V_{\text{on}})}{I_{\text{on}}} \tag{4.46b}$$

which is about 10 times longer than t_1'. The circuit in Fig. 4.10*b* results in a considerable speedup in the turn-off transient.

The circuits discussed so far have been *n*-channel devices. Normal *p*-channel devices always operate in the enhancement mode; that is, they have $V_T' < 0$. However, by ion implantation of *p*-type impurities one can make *p*-channel devices that have $V_T' > 0$. Such devices can thus be used as depletion-mode devices for the *p*-type version of Fig. 4.10*b*.

We finally observe that *n*-channel circuits are preferred to *p*-channel circuits. The reason is that the factor K is proportional to the mobility μ, and in silicon the electron mobility μ_n is considerably larger than the hole mobility μ_p. For silicon devices of comparable geometry and comparable bias, the *n*-channel device thus has a much larger current than the *p*-channel device, so switching is faster. For this reason *n*-channel technology is preferred.

4.5 DEVICE CAPACITANCES

In many MOS circuits one must take the effect of the device capacitances into account. If the substrate is connected to a separate supply voltage, one must distinguish four capacitances: the gate-source capacitance C_{gs}, the gate-drain capacitance C_{gd}, the substrate-source capacitance C_{bs}, and the substrate-drain capacitance C_{bd}. The direct capacitances between gate and substrate and between drain and source are usually negligible. One must also distinguish between the off and the on condition.

For the off condition there is no conducting channel, and C_{gs} and C_{gd} are relatively small because the gate only slightly overlaps the source and the drain. C_{bs} and C_{bd} are due to the source-substrate and the drain-substrate diodes.

For the on condition there is a conducting channel. Here the capacitance C_{gs} is mostly due to the oxide capacitance between the gate and the channel and is much larger than in the previous case. Also, the substrate-source capacitance is increased slightly by the capacitance between the substrate and the conducting channel. The equivalent circuit is shown in Fig. 4.11.

We now consider the circuit shown in Fig. 4.12. Let V_B be zero and V_A be high positive, and let the gate be so biased that there is no conducting channel. Then the voltage V_B stays this way until an appropriate gate pulse is applied, which turns the device on. When that happens, the load capacitance C_d presented by the inputs of subsequent FET devices is charged through the conducting channel until V_B reaches the value V_A. When the gate pulse terminates, V_B stays at the value V_A.

If a new pulse is applied and V_A has not changed, V_B does not change either. But if V_A has changed to a low voltage and the gate is then pulsed

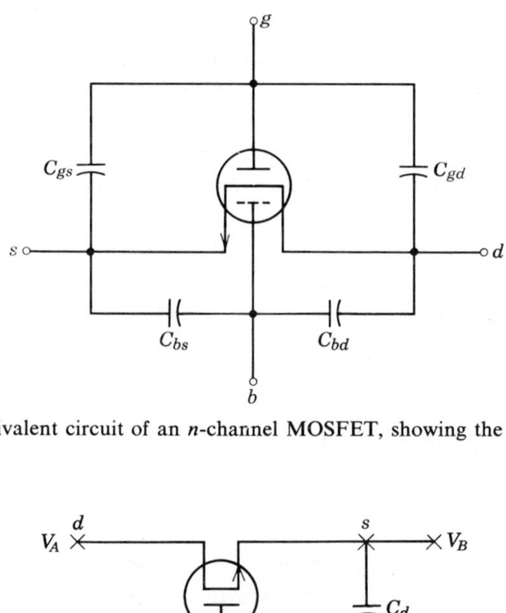

Figure 4.11. Equivalent circuit of an n-channel MOSFET, showing the four device capacitances.

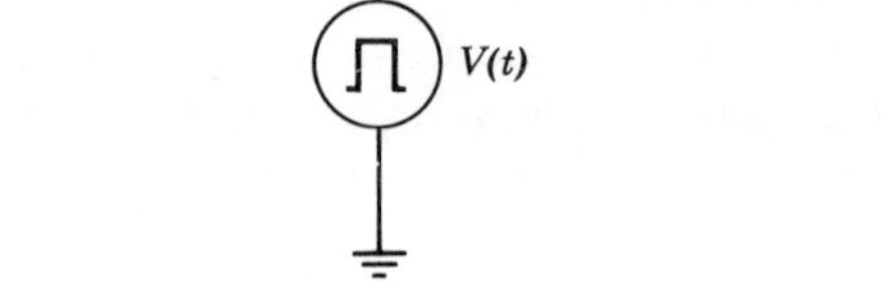

Figure 4.12. MOSFET circuit for use as a memory. A clock pulse turns the gate on and transmits the voltage V_A to the source.

afterward, current flows by reverse operation of the MOSFET until V_B reaches the value V_A. Therefore, if a pulse is applied to the gate periodically, V_B follows V_A but changes only on application of the next pulse after V_A has changed. This is important in clocked circuits such as those used in memory applications (Chapter 10).

PROBLEMS

4.1 The *n*-channel JFET in the circuit in Fig. 4.2*a* has $K = 1.0$ mA/V^2, $V_p = -2.3$ V, and a gate turn-on voltage of 0.70 V. The voltage V_i is chosen so that the gate draws current.

(a) If $V_{DD} = 20$ V and $R_d = 5000\ \Omega$, show that the device is in the ohmic state.
(b) Calculate R_{on}.
(c) Calculate V_0 approximately and accurately.

4.2 Repeat Problem 4.1 for a *p*-channel JFET having $K = 1.0$ mA/V^2, $V_p = +2.3$ V, turn-on voltage of the gate diode -0.70 V, $V_{DD} = -20$ V, and $R_d = 5000\ \Omega$.

4.3 In the circuit in Fig. 4.2*b*, $V_T = -2.0$ V, $V_i = 0.0$ V, $K = 0.2$ mA/V^2, $R_d = 5000\ \Omega$, and $V_{DD} = 10.0$ V. Find V_0.

4.4 Repeat Problem 4.3 if $V_T = +2.0$ V.

4.5 Repeat Problem 4.3 for a *p*-channel MOSFET having $V_T = -2.0$ V, $V_i = -6.0$ V, $K = 0.2$ mA/V^2, $R_d = 5000\ \Omega$, and $V_{DD} = -10$ V. Calculate V_0 both approximately and accurately.

4.6 Can logic circuits such as the one discussed in Problem 4.5 be connected in cascade and operate satisfactorily?

4.7 *p*-channel MOSFETs used in the logic circuit in Fig. 4.3 have $K_1 = 1.0$ mA/V^2, $K_2 = 0.16$ mA/V^2, $V_T = -3.0$ V, and $V_{DD} = -20$ V.

(a) Find the output voltage when the device is off, and the input voltage range for which the device is off.
(b) Find the output voltage for $V_i = -10$ V by first calculating I_{D0} and R_{on}.
(c) How accurate is this result? Find the accurate output voltage.

4.8 The circuit in Fig. 4.4 has $C_d = 0$, $C_{gs} = 5.0$ pF, $C_{dg} = 2.0$ pF, $g_m = 4.0 \times 10^{-3}$ mhos, and $R_d = 2500\ \Omega$.

(a) Find the input capacitance of the circuit including the Miller effect.
(b) Find the input time constant if $R_g = 2500\ \Omega$.

4.9 Calculate the turn-on time t_2 of a MOSFET circuit as in Fig. 4.4*a*, having $C_d = 20$ pF, $R_d = 5000\ \Omega$, and $V_{DD} = 20$ V; the MOSFET has $K = 0.5$ mA/V^2, $V_T = +2.0$ V, and $V(t) = V_1 = 12.0$ V. The turn-on time t_2 is again defined as the time needed for the transient to be 90% completed.

4.10 Calculate the turn-off time of the MOSFET circuit in Problem 4.9.

4.11 In the circuit in Fig. 4.8, $V_{DD} = 8$ V, $K_1 = K_2 = K = 1.0$ mA/V^2; $V_T = -2.0$ V for Q_p, $V_T = +2.0$ V for Q_n, and $C_0 = 20$ pF.

(a) Find $I_{D,\max}$ and R_{on} if $V_i = 8.0$ V.
(b) Replace the characteristic by a linear model and find the breakpoint V_{p0} between $I_D = V_D R_{\text{on}}$ and $I_D = I_{D,\max}$.
(c) Applying the piecewise linear approach discussed in Section 4.4c, calculate the turn-off time of the circuit, that is, the time required for V_0 to go from V_{DD} to $0.10 V_{DD}$.
(d) What is the turn-on time of the circuit?
(e) Compare the result thus obtained with what would be obtained with Eq. (4.37a) and comment on the difference.

CHAPTER

5 Bistable, Monostable, and Astable Multivibrators

This chapter deals with bistable, monostable and astable, multivibrators. A bistable multivibrator, also called a flip-flop circuit, is a feedback circuit with only resistors, which switches when triggered by a pulse (Section 5.1). A monostable multivibrator is a feedback circuit with one capacitor; it generates a complete waveform when triggered, and the duration of the waveform is determined by the capacitor (Section 5.2). An astable multivibrator is a feedback circuit with two capacitors, which has no stable operating point but generates a nonsinusoidal waveform, the period of which is determined by the capacitors (Section 5.3). Section 5.4 discusses an emitter-coupled Schmitt trigger which represents an alternate way of making flip-flops. Section 5.5 considers the operation of differential amplifiers which play a role in Schmitt triggers and other nonlinear circuits.

5.1 THE BISTABLE MULTIVIBRATOR

A *bistable multivibrator* or *flip-flop circuit* is a circuit consisting of two FETs or two transistors directly coupled together and operated in such a way that one device is off while the other one is on. This means that there are two stable operating conditions:

1. Device 1 is off; device 2 is on.
2. Device 1 is on; device 2 is off.

Hence the circuit is indeed a *bistable* circuit.

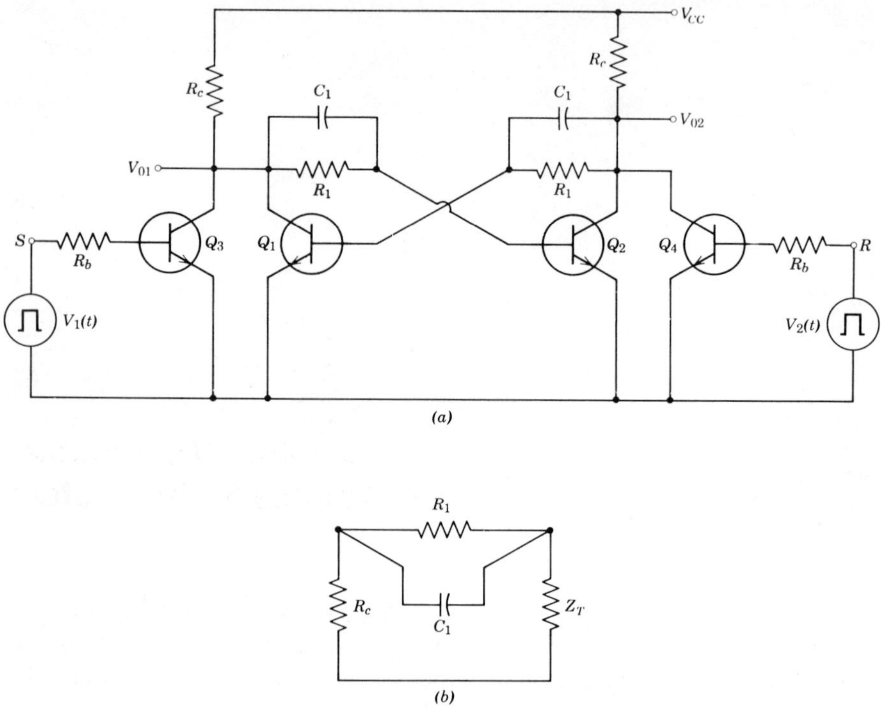

Figure 5.1. (*a*) Bistable transistor circuit with two transistor triggers. (*b*) Equivalent circuit used to explain the speedup in response by the capacitance C_1.

We now give several examples of such circuits involving either transistors or FETs.

5.1a The Transistor Bistable Multivibrator

Figure 5.1*a* shows a transistor flip-flop circuit. Here the resistors are chosen so that, when Q_1 is off, Q_2 is on and saturated, and vice versa. No negative voltage supply is needed, since $V_{CE,\text{sat}} < V_{BE0}$, so when one transistor is saturated the other one is off.

Let Q_1 be off; then

$$I_{B2} = \frac{V_{CC} - V_{BE0}}{R_c + R_1} \qquad I_{C2} = \frac{V_{CC} - V_{CE,\text{sat}}}{R_c} \tag{5.1}$$

Since Q_2 is saturated, R_c and R_1 must be chosen so that

$$I_{C2} < \beta_F I_{B2} \qquad \text{or} \qquad \beta_F > \frac{R_c + R_1}{R_c} \tag{5.1a}$$

since $V_{CC} - V_{CE,\text{sat}}$ and $V_{CC} - V_{BE0}$ differ only a little. This condition is basically not difficult to achieve, for R_1 and R_c can easily be made comparable. Often R_c and R_1 are of the order of 1000 Ω; this produces reasonably high currents. It also makes it possible to put the circuit in integrated form.

Example 1. In the circuit $R_1 = 1000\ \Omega$, $R_c = 500\ \Omega$, $V_{CC} = 4$ V, $V_{BE0} = 0.70$ V, $V_{CE,\text{sat}} \cong 0$ V, and $\beta_F = 50$. Find I_{B2} and I_{C2}.

Answer.

$$I_{B2} = \frac{4.0 - 0.7}{1500} = 2.2 \text{ mA} \qquad I_{C2} = \frac{4.0}{500} = 8 \text{ mA}$$

so

$$\beta_F > \frac{R_c + R_1}{R_c} = \frac{1500}{500} = 3$$

To switch the bistable circuit from one stable condition to the other, a trigger pulse must be applied. Figure 5.1*a* shows how this can be done with two transistor triggers. A positive pulse applied to transistor Q_3 reduces V_{CE1} to 0 V, hence turns Q_2 off so that it gives a 1 output. A positive pulse applied to transistor Q_4 reduces V_{CE2} to 0 V, hence turns Q_1 off so that it gives a 1 output. Hence the pulse applied to the output terminals of Q_3 gives a 1 output at terminal 2, and the pulse applied to the input terminals of Q_4 gives a 0 output at terminal 2. For these reasons terminal S is called the *set terminal*, since it sets the output of terminal 2 at 1, and terminal R is called the *reset terminal*, since it sets the output of terminal 2 at 0.

The value of the voltage V_{02} at terminal 2 for a 0 output is $V_{CE,\text{sat}}$, and the value of V_{02} for a 1 output is

$$V_{02} = (V_{CC} - V_{BE0})\frac{R_1}{R_1 + R_c} + V_{BE0} \tag{5.2}$$

To make V_{02} large, R_1 should be considerably larger than R_c (e.g., by a factor of 3 to 10).

It should be observed that $V_{BE0} > V_{CE,\text{sat}}$. This means that Q_1 is already off before Q_2 is saturated. There may be a possibility that both transistors are on and nonsaturated during the switching process, but this is not a stable configuration. Let us assume that Q_1 and Q_2 are both on and nonsaturated and that I_{C1} increases by an amount ΔI_{C1}. Then V_{C1} decreases, which causes a decrease in I_{B2}, hence in I_{C2}. This raises V_{C2}, hence increases I_{B1}, which increases I_{C1} still further. If the total gain of the circuit is larger than unity, this will continue until Q_1 is fully on and Q_2 fully off. In the same way a small decrease in I_{C1} is amplified and ultimately turns Q_1 off and Q_2 on.

To speed up the response, capacitances C_1 are connected in parallel with resistances R_1, and are chosen so that

$$C_1 R_1 = \tau_T \tag{5.3}$$

where τ_T takes account of the Miller effect. If Q_1 is off, the equivalent circuit "seen" by the base of Q_2 is as shown in Fig. 5.1*b*. When the base current is sufficiently large, the effect of the base impedance Z_T is small and can be neglected. The time constant of the circuit is then

$$\tau' = \frac{R_1 R_c}{R_1 + R_c} C_1 = \frac{R_c}{R_1 + R_c} \tau_T \tag{5.4}$$

This gives an idea of the degree of speedup that can be achieved.

Example 2. If in Example 1 we have $\tau_T = 5 \times 10^{-8}$ s, find τ'.

Answer.

$$\tau' = \frac{500}{1500} 5 \times 10^{-8} = 1.67 \times 10^{-8} \text{ s}$$

In this circuit the flip-flop is made by cross-coupling two inverters. Since integrated nand and nor gates (Chapter 6) are commercially available in package form, it is common practice to make flip-flop circuits by cross-coupling two nand or two nor gates (Chapter 10).

5.1b The JFET Bistable Multivibrator

As a second example we discuss the JFET flip-flop. Figure 5.2*a* shows the circuit, involving two JFETs with an *n*-type channel. It is necessary to use two batteries, one with positive polarity (V_{DD}) and one with negative polarity (V_{GG}). The reason is that the FETs have a pinch-off voltage, $V_p < 0$, so a negative gate voltage is needed to turn the device off and a positive drain voltage is needed to keep it on. One of the main reasons for discussing this circuit is to demonstrate how devices with negative pinch-off (or turn-on) voltages can be used.

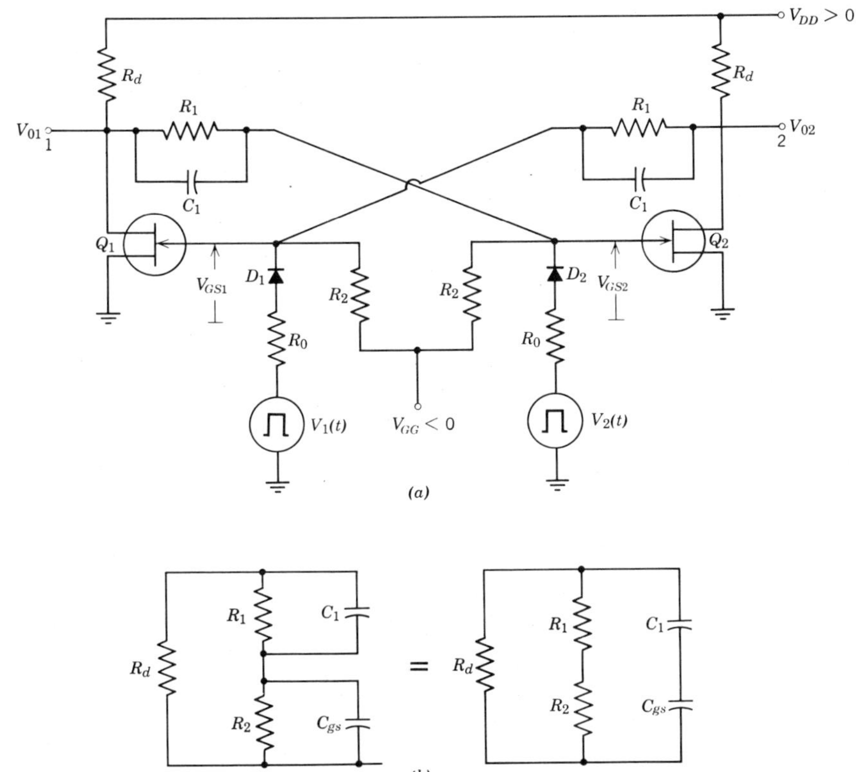

Figure 5.2. (*a*) Bistable JFET circuit with two diode triggers. (*b*) Equivalent circuit used to explain the speedup in response by the capacitance C_1.

We now assume that Q_1 is off and Q_2 on, and that the circuit has been designed so that $V_{DS2} = V_{\text{on}}$ is close to zero, or Q_2 operates in the ohmic regime. We then have, for V_{GS2},

$$V_{GS2} = V_{DD}\frac{R_2}{R_d + R_1 + R_2} + V_{GG}\frac{R_d + R_1}{R_d + R_1 + R_2} > V_p \tag{5.5}$$

This is explained as follows. Since Q_1 is off, the supply for G_2 is a simple potentiometer circuit, provided that the gate of Q_2 does not draw too much current. This is certainly the case if $V_{GS2} < 0.55$ V. The required range of V_{GS2} is therefore

$$V_p < V_{GS2} < 0.55 \text{ V} \tag{5.5a}$$

with V_{GS2} preferably close to the highest value.

For Q_1, however,

$$V_{GS1} = V_{DS2}\frac{R_2}{R_1+R_2} + V_{GG}\frac{R_1}{R_1+R_2} < V_p \tag{5.6}$$

Here $V_{DS2} = V_{on}$ is developed over the resistance R_{on} (compare Chapter 4), which is small in comparison with $R_1 + R_2$. Since V_{DS2} is small, Eq. (5.6) can be approximated as

$$V_{GG} < V_p\frac{R_1+R_2}{R_1} \tag{5.6a}$$

Both conditions (5.5) and (5.6) must be satisfied simultaneously. The other stable condition is found by interchanging the subscripts 1 and 2 in V_{GS1}, V_{GS2}, V_{DS1}, and V_{DS2}.

If Q_2 operates in the ohmic regime, the output voltage V_{02} of terminal 2 is V_{on} which is close to 0 V. If Q_2 is off, however, and Q_1 is drawing gate current, $V_{GS1} \cong V_{GS0} = 0.70$ V and

$$V_{02} = (V_{DD} - V_{GS0})\frac{R_1}{R_1+R_d} + V_{GS0} \tag{5.7}$$

So it is important to make R_1 considerable larger than R_d (e.g., by a factor of 3 to 10); otherwise V_{02} becomes too low.

To switch the bistable circuit from one stable condition to the other, a trigger pulse must be applied. Figure 5.2*a* shows how this can be done with two diode triggers. The resistors R_0 are needed to protect the gates and the diodes.

We assume that only one of the devices is on at the same time. During switching both devices may be on at the same time but, as in the transistor case, the situation $V_{GS1} > V_p$ and $V_{GS2} > V_p$ is unstable if the total gain of the circuit is larger than unity. Let V_{GS1} increase by an amount ΔV_{GS1}; this is amplified in Q_1 and fed into the gate of Q_2, where it decreases V_{GS2}. This change in V_{GS2} is amplified in Q_2 and fed into the gate of Q_1, where it enhances the original increase ΔV_{GS1}. If the overall gain of the system is larger than unity, the original variation will grow until Q_1 is fully on and Q_2 is fully off. Conversely, if V_{GS1} decreases by an amount ΔV_{GS1}, this increases V_{GS2}, which in turn decreases V_{GS1}, and this will continue until Q_1 is fully off and Q_2 fully on.

Example 3. $V_{DD} = 30$ V, $V_{GG} = -30$ V, $V_p = -3$ V, $R_{on} = 300\ \Omega$, $R_d = 5000\ \Omega$, $R_1 = 10{,}000\ \Omega$, and $R_2 = 15{,}000\ \Omega$. Find V_{GS2} and V_{GS1}.

Answer.

$$V_{GS2}=30\frac{15}{5+10+15}-30\frac{5+10}{5+10+15}=15-15=0 \text{ V}$$

well satisfying the condition $V_p < V_{GS2} < 0.55$ V.

$$V_{\text{on}}=V_{DD}\frac{R_{\text{on}}}{R_d+R_{\text{on}}}=30\frac{300}{5300}=1.7 \text{ V}$$

$$V_{GS1}=1.7\frac{15}{10+15}-30.0\frac{10}{10+15}=1.0-12.0=-11.0 \text{ V}$$

well satisfying the condition $V_{GS1} < V_p$.

The input capacitance C_{gs} of the FET shunts R_2, and this slows the switching. The time constant involved in the switching is, if we neglect the Miller effect,

$$\tau=C_{gs}\frac{R_2(R_1+R_d)}{R_2+R_1+R_d} \tag{5.8}$$

The switching can be speeded up by connecting a capacitance C_1 in parallel with the resistance R_1, so that

$$R_1C_1=R_2C_{gs} \tag{5.9}$$

The time constant τ' is given by the equivalent circuit in Fig. 5.2*b* as

$$\tau'=\frac{R_d(R_1+R_2)}{R_d+R_1+R_2}\frac{C_1C_{gs}}{C_1+C_{gs}} \tag{5.9a}$$

Since $R_1C_1=R_2C_{gs}$, the series connection between R_1, C_1 and R_2, C_{gs} corresponds to the parallel connection between R_1+R_2 and C', where C' is the capacitance of the series connection between C_1 and C_{gs}. This speeds up the response.

Actually, the situation is slightly more complicated, since the capacitance C_{dg} between the drain and the gate gives rise to the Miller effect, which increases the input capacitance above C_{gs}. Hence C_1 should be somewhat larger than Eq. (5.9a) indicates.

Example 4. If $C_{gs}=5.0$ pF, find the value of C_1 needed for a speedup of the response, and the time constants τ and τ' for Example 3.

Answer.

$$\tau = 5\times 10^{-12}\frac{15\times 15}{30}10^3 = 37.5\times 10^{-9}\ \text{s}$$

$$C_1 = \frac{C_{gs}R_2}{R_1} = \frac{5\times 10^{-12}\times 15}{10} = 7.5\ \text{pF}$$

$$\tau' = \frac{5\times 25}{30}10^3\times\frac{7.5\times 5.0}{12.5}10^{-12} = 12.5\times 10^{-9}\ \text{s}$$

5.1c The MOSFET Bistable Multivibrator with Enhancement-Mode Devices

MOSFETs operating in the depletion mode, such as n-channel MOSFETs with $V_T < 0$, operate in the same way as JFET circuits, hence no further discussion is needed. The only difference is that the gate never draws current, so there is no restriction on V_{GS} when a device is on.

We saw in Chapter 4 that enhancement-mode devices permitted the development of circuits without resistors. Two such amplifiers directly connected can produce a simple bistable circuit which is easily put in integrated form. If the gain per stage is large enough, the circuit works with a single battery.

Figure 5.3 shows such a circuit with p-channel MOSFETs. Assume that all four devices have the same turn-on voltage, $V_T < 0$. Assume further that Q_1 and Q_2 have a K factor K_1, and L_1 and L_2 have a K factor K_2, where $K_1 > K_2$, since each half must give an amplification, or $(K_1/K_2)^{1/2} > 1$. We shall see in a moment what value of K_1/K_2 is required.

Let Q_1 be off, or $V_{GS1} > V_T$, and Q_2 on, or $V_{GS2} < V_T$; then

$$V_{GS2} = V_{DS1} = V_{DD} - V_T \qquad V_{GS1} = V_{DS2} = V_{\text{on}} > V_T \tag{5.10}$$

Because Q_2 is on, and $V_{DS2} = V_{\text{on}} \cong 0$, L_2 has a current

$$I_{DL2} = -K_2(V_{DD} - V_T - V_{DS2})^2 \cong -K_2(V_{DD} - V_T)^2 \tag{5.11}$$

and Q_2 has a current

$$I_{D2} = -K_1\left[2(V_{GS2} - V_T)V_{DS2} - V_{DS2}^2\right] \cong -2K_1(V_{GS2} - V_T)V_{DS2} \tag{5.12}$$

But since the devices are in series, $I_{DL2} = I_{D2}$, so

$$K_2(V_{DD} - V_T)^2 = 2K_1(V_{GS2} - V_T)V_{DS2} = 2K_1(V_{DD} - 2V_T)V_{DS2}$$

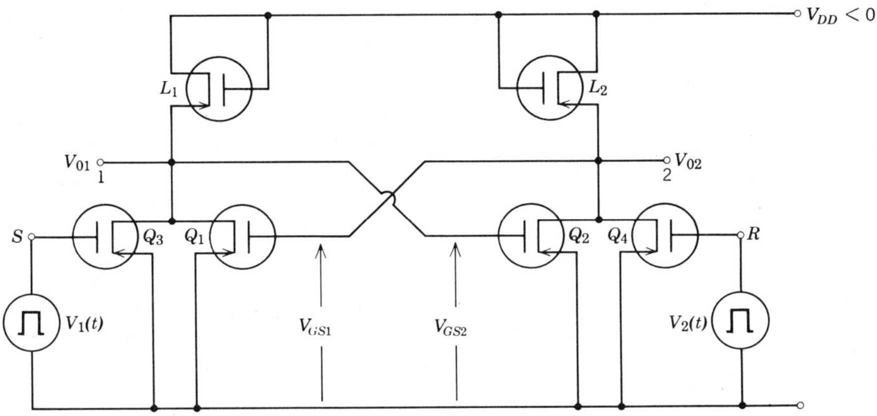

Figure 5.3. Bistable MOSFET circuit with two MOSFET triggers. All MOSFETs are enhancement-mode p-channel devices.

or

$$V_{DS2} = \frac{K_2 (V_{DD} - V_T)^2}{2K_1 (V_{DD} - 2V_T)} \tag{5.13}$$

since $V_{GS2} = V_{DD} - V_T$. But according to Eq. (5.10) $V_{GS1} = V_{DS2} > V_T$, which leads to the requirement

$$\frac{K_2 (V_{DD} - V_T)^2}{2K_1 (V_{DD} - 2V_T)} > V_T \qquad \text{or} \qquad \frac{K_1}{K_2} > \frac{(V_{DD} - V_T)^2}{2(V_{DD} - 2V_T) V_T} \tag{5.14}$$

(Note that $V_{DD} - 2V_T$ and V_T are negative quantities.) The circuit works properly if this condition is satisfied.

It should be understood from Section 4.3b that the expression for V_{DS2} is slightly inaccurate. Applying Eq. (4.27) and substituting $V_i = V_{DD} - V_T$ yields*

$$V_{DS2} = \frac{K_1 (V_{DD} - 2V_T) + K_2 (V_{DD} - V_T)}{K_1 + K_2} + \left\{ \left[\frac{K_1 (V_{DD} - 2V_T) + K_2 (V_{DD} - V_T)}{K_1 + K_2} \right]^2 - \frac{K_2 (V_{DD} - V_T)^2}{K_1 + K_2} \right\}^{1/2} \tag{5.15}$$

*We need the plus sign, since otherwise Q_1 would not be in non-pinch-off condition.

Triggering the circuit to effect the switching from the one stable state to the other can be done in the same way as for the JFET. The only difference is that because p-type devices are used, the polarity of the diodes and of the triggering pulse must be reversed. Since diode triggers are not so easily put on the same chip with MOSFETs, Fig. 5.3 shows two MOSFET triggers driven by negative pulses. Input S (set input), when triggered, sets the output V_{02} at $V_{DD}-V_T$, which is a 1, whereas input R (reset input), when triggered, sets the output V_{02} at V_{on}, which is a 0.

Example 5. If $V_{DD}=-20$ V and $V_T=-4$ V, find the inequality for K_1/K_2.

Answer.

$$\frac{K_1}{K_2} > \frac{(-20+4)^2}{2(-20+8)(-4)} = \frac{16^2}{2\times 12\times 4} = 2\tfrac{2}{3}$$

which is not difficult to satisfy.

Example 6. Compare Eq. (5.13) with the more accurate result from Eq. (5.15) if $V_{DD}=-20$ V, $V_T=-4$ V, and $K_1/K_2=4$.

Answer. Equation (5.13) yields

$$V_{DS2} = -\frac{1}{8}\frac{(16)^2}{12} = -2.67 \text{ V}$$

Equation (5.15) gives

$$V_{DS2} = \frac{-12-0.25\times 16}{1.25} + \left[\left(\frac{-12-0.25\times 16}{1.25}\right)^2 - \frac{0.25\times 16^2}{1.25}\right]^{1/2}$$

$$= -12.80 + \left[(12.80)^2 - 51.20\right]^{1/2} = -12.80 + (112.64)^{1/2} = -2.19 \text{ V}$$

The difference is significant; the result from Eq. (5.13) is somewhat too low.

5.2 THE MONOSTABLE MULTIVIBRATOR

A monostable multivibrator is a circuit that goes through a complete waveform when triggered. We now discuss two examples.

5.2a The Transistor Monostable Multivibrator

By replacing the one R_1C_1 combination in Fig. 5.1a by a capacitor C and returning the floating base to V_{CC} via a resistor R, we obtain the circuit in Fig. 5.4. The resistors R and R_c are chosen so that Q_2 is normally on and saturated, whereas Q_1 is normally off. If Q_1 is triggered by a pulse, it is turned on, which turns Q_2 off. The base of Q_2 is then charged through the resistor R until its voltage reaches the value V_{BE0}, at which point Q_2 is again turned on. The resistor R_0 is needed to limit the current.

If the transistor Q_1 is triggered at $t=0$, just before triggering,

$$V_{01}(0^-)=V_{CC} \qquad V_{B2}(0^-)=V_{BE0}\cong 0.70 \text{ V} \tag{5.16}$$

Right after triggering and switching, that is, at $t=0^+$,

$$V_{01}(0^+)=V_{CE,\text{sat}} \qquad \text{hence} \qquad V_{B2}(0^+)=V_{BE0}-V_{CC}+V_{CE,\text{sat}} \tag{5.17}$$

since the change in $V_{C1}(t)$ at $t=0$ cannot be taken up by the capacitor C but must be transmitted to B_2. We assume that B_2 can withstan.' this negative voltage without breaking down.

Since the capacitor C is charged through the resistance R, the time constant is RC, and $V_{B2}(t)$ will rise to the final value $V_{B2}(\infty)=V_{CC}$ if

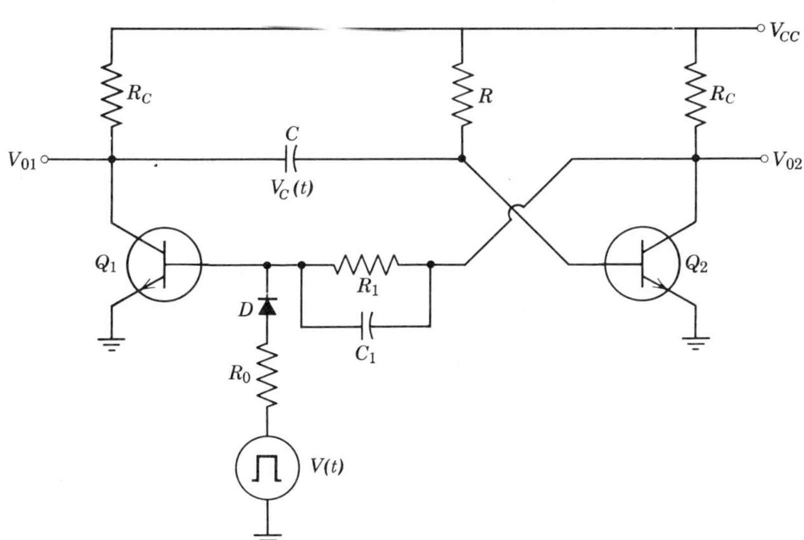

Figure 5.4. Monostable transistor circuit with a diode trigger.

nothing intervenes. Consequently (see Chapter 1), if $\tau_s = RC$,

$$V_{B2}(t) = V_{B2}(0^+)\exp\left(-\frac{t}{RC}\right) + V_{B2}(\infty)\left[1-\exp\left(-\frac{t}{RC}\right)\right]$$

$$= V_{CC} - (2V_{CC} - V_{BE0} - V_{CE,\text{sat}})\exp\left(-\frac{t}{RC}\right) \tag{5.18}$$

This part of the waveform stops if $V_{B2}(t)$ reaches the value V_{BE0}; at that point Q_2 is turned on, and Q_1 is turned off. Let this occur at $t = T_s$; then

$$\exp\left(-\frac{T_s}{RC}\right) = \frac{V_{CC} - V_{BE0}}{2V_{CC} - V_{BE0} - V_{CE,\text{sat}}} \qquad T_s = RC\ln\left(\frac{2V_{CC} - V_{BE0} - V_{CE,\text{sat}}}{V_{CC} - V_{BE0}}\right) \tag{5.19}$$

For $V_{CC} \gg V_{BE0} + V_{CE,\text{sat}}$ this reduces to

$$T_s \cong RC\ln 2 = 0.69RC \tag{5.19a}$$

Usually it is better to use the more accurate Eq. (5.19).

The original situation has not yet been fully restored, however, since V_{C1} has not yet recovered. It takes some time before this is accomplished, which has a bearing on the repetition rate with which the circuit can be triggered.

To investigate this problem, we observe that, just before switching, that is, at $t = T_s^-$, Q_1 is still saturated, or $V_{C1}(T_s^-) = V_{CE,\text{sat}} \cong 0$, and $V_{B2}(T_s^-) = V_{BE0} \cong 0.70$ V. Right after switching, that is, at $t = T_s^+$, V_{B2} has hardly changed, hence $V_{C1}(T_s^+)$ remains at about 0 V. Since $V_{B2}(t)$ remains fixed at V_{BE0}, the capacitor C must charge through the resistance R_c, making $V_{C1}(t)$ approach the ultimate value $V_{C1}(\infty) = V_{CC}$. Since the time constant $\tau = CR_c$, and $V_{C1}(T_s^+) \cong 0$,

$$V_{C1}(t) = V_{CC}\left[1-\exp\left(-\frac{t-T_s}{CR_c}\right)\right] \tag{5.20}$$

(see Chapter 1).

In this case it is not sufficient to have the transient 90% complete, we must wait somewhat longer. A good compromise for the recovery time $T_R = t - T_s$ is

$$T_R \cong 4R_cC \tag{5.20a}$$

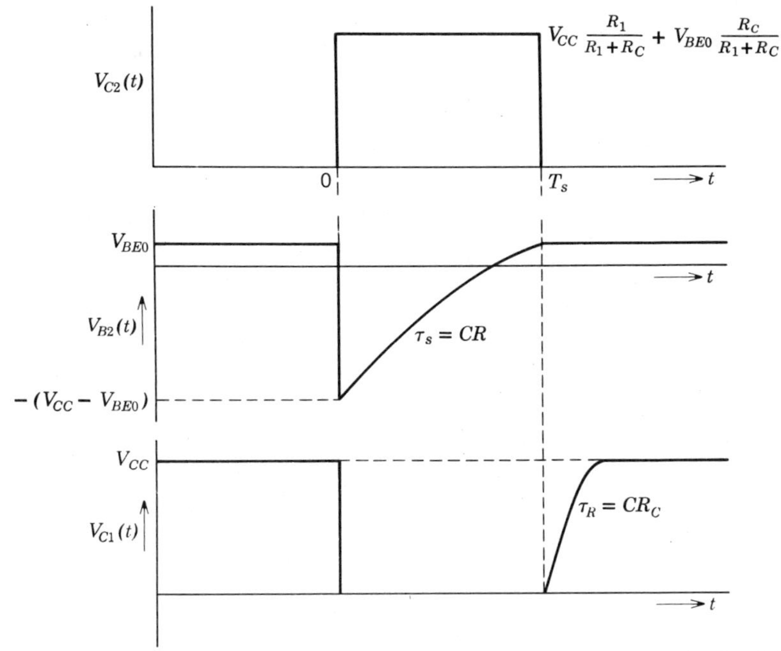

Figure 5.5. $V_{C2}(t)$, $V_{B2}(t)$, and $V_{C1}(t)$ as a function of time for the circuit in Fig. 5.4. We assume here $V_{CE,\text{sat}}=0$.

If we want T_R to be smaller than the pulse duration T_s, we must require

$$4R_cC < RC\ln\left(\frac{2V_{CC}-V_{BE0}-V_{CE,\text{sat}}}{V_{CC}-V_{BE0}}\right) \tag{5.20b}$$

or, if $V_{CC} \gg V_{BE0}$,

$$R > \frac{4R_c}{\ln 2} = 5.8R_c \tag{5.20c}$$

We cannot make the resistance R too large, however, otherwise the transistor would not be saturated. In analogy with Eq. (5.1a) this requires

$$\beta_F > \frac{R_c+R}{R_c} \qquad \text{or} \qquad R < \beta_F R_c \tag{5.21}$$

so the requirements for R are approximately

$$5.8R_c < R < \beta_F R_c \tag{5.21a}$$

Therefore $R \cong 10R_c$ is quite sufficient for high-β transistors. These considerations are also important for astable multivibrators (Section 5.3a).

Figure 5.5 shows the waveform of $V_{C2}(t)$, $V_{B2}(t)$, and $V_{C1}(t)$ if $V_{CE,\text{sat}}=0$.

Example 1. If $R = 10{,}000\ \Omega$, $R_c = 1000\ \Omega$, and $C = 200$ pF, find T_s and T_R, and determine whether or not the condition $T_R < T_s$ is satisfied.

Answer. $T_s = 0.69RC = 0.693 \times 10^4 \times 2 \times 10^{-10} = 1.39\ \mu$s. $T_R = 4R_cC = 4 \times 10^3 \times 2 \times 10^{-10} = 0.80\ \mu$s, or $T_R < T_s$.

Thus far we have assumed that the base of the transistor Q_2 can withstand the back voltage $V_{B2}(0^+) = V_{BE0} - V_{CC}$. Often this is not the case, and the emitter-base junction breaks down at the voltage $-BV_{BE0}$ given in device manuals. This breakdown affects the waveform and the duration of the pulse in the output. It is not difficult to take this into account, and we do so here in order to demonstrate how junction breakdown problems must be handled.

We shorten BV_{BE0} to V_{B0}. If Q_1 is turned on at $t = 0$, then $V_{B2}(0^+)$ cannot drop below $-V_{B0}$. Since $V_c(0^+) = V_{CC} - V_{BE0}$ is the voltage drop across the capacitor, we have $V_{c1}(0^+) = V_c(0^+) - V_{B2}(0^+) = V_{CC} - V_{BE0} - V_{B0}$. Consequently, Q_1 is not driven to saturation immediately; its currents are

$$I_{B1} = \frac{V_{CC} - V_{BE0}}{R_c + R_1} \qquad I_{C1} = \beta_F I_{B1}$$

and the capacitance C is discharged by the current $I_{C1} - [V_{CC} - V_{C1}(t)]/R_c$. If the last term is neglected, since it is small, the discharge is completed at $t = T_s'$, where

$$T_s' = \frac{V_{C1}(0^+)C}{\beta_F I_{B1}} = \frac{C(R_c + R_1)}{\beta_F}\,\frac{V_{CC} - V_{BE0} - V_{B0}}{V_{CC} - V_{BE0}} \tag{5.22}$$

This discharge mode of the capacitor C stops when the breakdown current of the base of Q_2 stops. The value of $V_{C1}(t)$ at $t = T_s'$ is equal to $V_{CE,\text{sat}} \cong 0$. Since $V_{B2}(T_s{}^+) = -V_{B0}$ and $V_{B2}(\infty) = V_{CC}$, we have for $t > T_s'$, because the capacitor is now charged through the resistor R,

$$V_{B2}(t) = -V_{B0}\exp\left(-\frac{t - T_s'}{CR}\right) + V_{CC}\left[1 - \exp\left(-\frac{t - T_s'}{CR}\right)\right] \tag{5.23}$$

This part of the transient stops if $V_{B2}(t) = V_{BE0}$; at that time Q_2 is turned on and Q_1 is turned off. Let this be at $t - T_s' = T_s''$; then

$$\exp\left(-\frac{T_s''}{RC}\right) = \frac{V_{CC} - V_{BE0}}{V_{CC} + V_{B0}} \qquad \text{or} \qquad T_s'' = RC\ln\left(\frac{V_{CC} + V_{B0}}{V_{CC} - V_{BE0}}\right) \tag{5.23a}$$

Hence the total time T_s of the waveform is

$$T_s = T_s' + T_s'' \tag{5.23b}$$

Example 2. If $C=200$ pF, $V_{CC}=6.0$ V, $BV_{BEO}=3.5$ V, $V_{BEO}=0.70$ V, $V_{CE,\text{sat}}=0.00$ V, $\beta_F=50$, $R_1=2000$ Ω, $R=10{,}000$ Ω, and $R_c=1000$ Ω, find T_s' and T_s''.

Answer.

$$T_s' = \frac{2\times 10^{-10}\times 3000}{50}\times\frac{1.8}{5.3} = 0.41\times 10^{-8}\ \text{s}$$

$$T_s'' = 2\times 10^{-10}\times 10^4 \ln\left(\frac{9.5}{5.3}\right) = 1.17\times 10^{-6}\ \text{s}$$

The waveform for $V_{C2}(t)$ has not changed; $V_{B2}(t) = -V_{BO}$ for $0<t<T_s'$; $V_{C1}(0^+) = V_{CC} - V_{BEO} - V_{BO}$ and goes linearly to zero for $0<t<T_s'$, reaching zero at $t=T_s'$. It is not necessary to draw a graph. We also see that again $T_R < T_s$.

5.2b The MOSFET Monostable Multivibrator

Figure 5.6 shows a monostable multivibrator involving p-channel enhancement-mode MOSFETs. Device Q_2 is normally on and operating in the

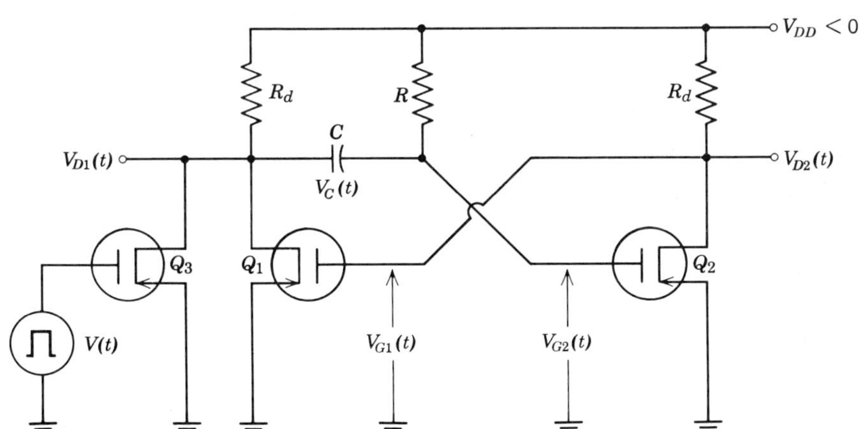

Figure 5.6. Monostable MOSFET circuit with a MOSFET trigger. All devices are enhancement-mode p-channel devices.

ohmic regime, so Q_1 is off. By triggering Q_1 one can turn Q_1 on temporarily, until the stable configuration, Q_2 on and Q_1 off, is restored. The triggering is done with a MOSFET Q_3.

Let Q_1 be triggered at $t=0$. Just before triggering, that is, at $t=0^-$,

$$V_{G2}(0^-)=V_{DD} \qquad V_{D1}(0^-)=V_{DD} \qquad V_C(0^-)=0 \tag{5.24}$$

where $V_C(t)$ is the voltage across the capacitor C. After triggering and completed switching, that is, at $t=0^+$,

$$V_{D1}(0^+)=V_{G2}(0^+)=V_{\text{on}}>V_T \tag{5.24a}$$

and $V_{D1}(t)=V_{\text{on}}$ is fixed from then on. Therefore the capacitor charges through the resistor R, and $V_{G2}(t)$ will decrease to the final value $V_{G2}(\infty) = V_{DD}$ if nothing intervenes. Hence the time constant of the transient is $\tau_s = CR$ and (see Chapter 1)

$$V_{G2}(t)=V_{G2}(0^+)\exp\left(-\frac{t}{CR}\right)+V_{G2}(\infty)\left[1-\exp\left(-\frac{t}{RC}\right)\right]$$

$$=V_{DD}-(V_{DD}-V_{\text{on}})\exp\left(-\frac{t}{RC}\right) \tag{5.25}$$

This part of the transient stops when $V_{G2}(t)$ reaches the value V_T. Let this be the case at $t=T_s$; then

$$\exp\left(-\frac{T_s}{RC}\right)=\frac{V_T-V_{DD}}{V_{\text{on}}-V_{DD}} \qquad T_s=RC\ln\left(\frac{V_{\text{on}}-V_{DD}}{V_T-V_{DD}}\right) \tag{5.26}$$

This is a relatively short time interval (see Example 3).

We have, just before switching, that is, at T_s^-, $V_{D1}(T_s^-)=V_{\text{on}}$ and $V_{G2}(T_s^-)=V_T$, so $V_C(T_s^-)=V_{\text{on}}-V_T$. The original situation has therefore not been fully restored, since the charge on the capacitor C needs to be removed. Since the discharging of C occurs through $R+R_d$, the recovery part of the transient has a time constant

$$\tau_R=C(R+R_d) \tag{5.27}$$

To speed up the recovery we require that τ_R be not much larger than T_s, or $R_d<R$. The full recovery time T_R is then a few times τ_R. This is basically not difficult, but the condition $T_R<T_s$ cannot be satisfied. Figure 5.7 shows the waveform of $V_{D2}(t)$.

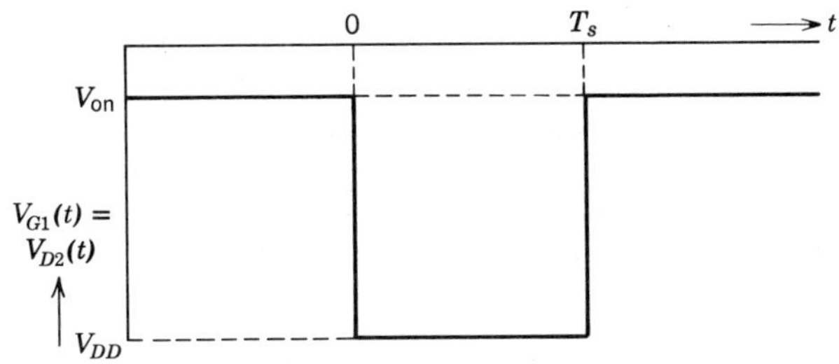

Figure 5.7. $V_{G1}(t) = V_{D2}(t)$ as a function of time for the circuit in Fig. 5.6.

Example 3. If $V_{DD} = -20.0$ V, $V_T = -4.0$ V, $R = 10{,}000\ \Omega$, $R_d = 5000\ \Omega$, $C = 1000$ pF, and $V_{on} = -1.0$ V, find τ_s, τ_R, T_s, and T_R.

Answer. $\tau_s = CR = 10^{-9} \times 10^4 = 10^{-5}$ s. $\tau_R = C(R + R_d) = 10^{-9} \times 1.5 \times 10^4 = 1.5 \times 10^{-5}$ s.

$$T_s = \tau_s \ln\left(\frac{V_{on} - V_{DD}}{V_T - V_{DD}}\right) = 10^{-5} \ln\left(\frac{19}{16}\right) = 1.7 \times 10^{-6} \text{ s}$$

$$T_R \cong 4\tau_r = 6 \times 10^{-5} \text{ s} \gg T_s$$

The condition $T_R < T_s$ is therefore not satisfied.

5.3 THE ASTABLE MULTIVIBRATOR

An astable multivibrator is a circuit that generates a periodic non-sinusoidal signal and does not need triggering. It can be made by modifying a monostable circuit. We discuss a few examples in greater detail.

5.3a The Astable Transistor Multivibrator

Figure 5.8 shows an astable transistor multivibrator circuit. It contains two equal capacitors C_1 and C_2 to connect one collector with the other base, and equal resistors R_1 and R_2 to feed the two bases, so the time constants are equal:

$$C_1R_1 = C_2R_2 = CR = \tau \qquad (5.28)$$

Since the theory of this circuit is identical to the one given for the monostable circuit in Section 5.2a, we can apply the results from that

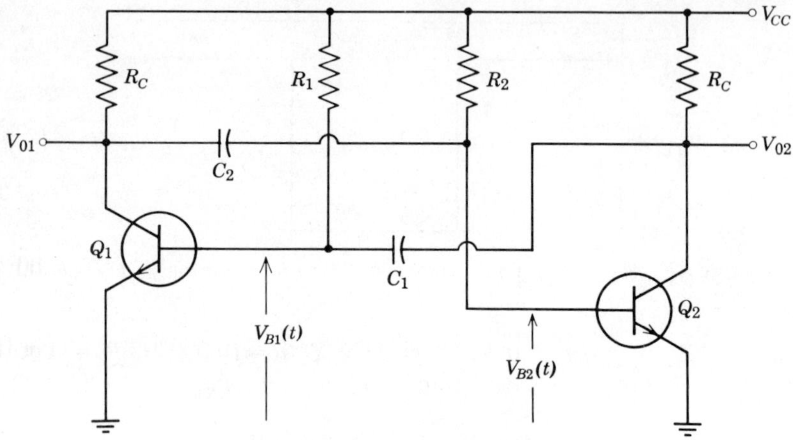

Figure 5.8. Astable transistor circuit.

section directly. We require that the transistor that is on be saturated and that the recovery time T_R of the circuit be shorter than the duration of the pulse T_s. According to Eq. (5.21a), this means

$$5.8R_c < R < \beta_F R_c \tag{5.29}$$

which is easily met in practice. Moreover, the total period T of the waveform is $2T_s$, since T_s is the time that one device is on, and the full period includes an on period for each device. Therefore

$$T = 2T_s = 2RC\ln\left(\frac{2V_{CC} - V_{BE0} - V_{CE,\text{sat}}}{V_{CC} - V_{BE0}}\right) \qquad \text{or} \qquad T \cong 2CR\ln 2 = 1.38CR \tag{5.30}$$

Usually the more accurate expression given by the first half of Eq. (5.30) is needed.

Example 1. If $R = 10{,}000\ \Omega$, $R_c = 1000\ \Omega$, $\beta_F = 50$, and $C = 1000$ pF, determine whether or not the condition (5.29) is satisfied and calculate the full period T of the waveform.

Answer. Equation (5.29) becomes $5.8\times10^3 < R < 50\times10^3$, and this is true. Since $\tau = CR = 10^{-9}\times10^4 = 10^{-5}$ s, $T = 1.38\tau = 13.8\ \mu$s.

Figure 5.9 shows the complete waveform for a full cycle for $V_{02}(t)$ and $V_{B2}(t)$ if $V_{CE,\text{sat}} = 0$.

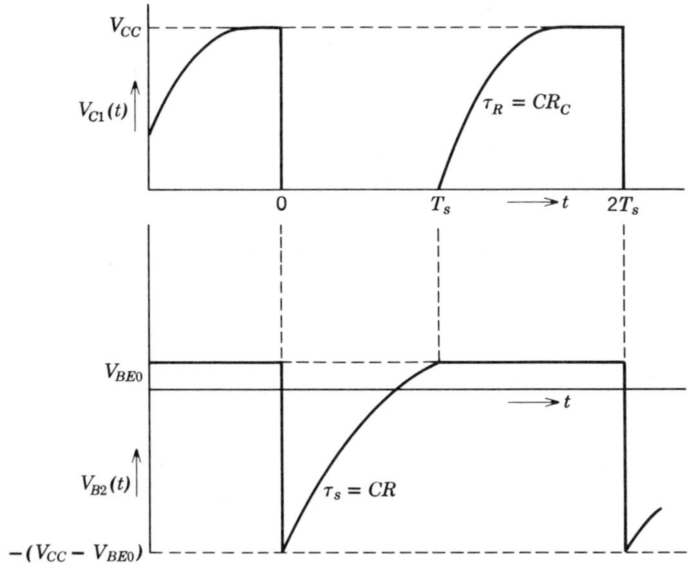

Figure 5.9. $V_{C1}(t)$ and $V_{B2}(t)$ as functions of time for the circuit in Fig. 5.8.

5.3b The Astable JFET Multivibrator

The circuit in Fig. 5.6 has the disadvantage of a relatively short pulse time and a relatively long recovery time. If such a monostable circuit is modified to make it an astable circuit, the waveform is not as easily evaluated. We therefore discuss a JFET circuit that does not have this defect.

Figure 5.10 shows the circuit. We assume $C_1 = C_2 = C$ and $R_{g1} = R_{g2} = R_g$. Let V_p be the pinch-off voltage of the JFETs. We assume that at $t=0$ the circuit switches from Q_1 off, Q_2 on to Q_1 on, Q_2 off.

Just before switching, that is, at $t=0^-$, we have

$$V_{D1}(0^-) = V_{DD} \qquad V_{G1}(0^-) = V_p \qquad V_{G2}(0^-) = V_{GS0} \cong 0.70 \text{ V} \tag{5.31}$$

since the device that is on is drawing a gate current $V_{DD} - 0.70) / R_{g2}$. Just after switching, that is, at $t=0^+$, we have instead

$$V_{D1}(0^+) = V_{\text{on}} \qquad V_{G1}(0^+) = V_{GS0} \cong 0.70 \text{ V} \tag{5.32}$$

Hence, since V_{D1} shifts by an amount $V_{DD} - V_{\text{on}}$ during switching,

$$V_{G2}(0^+) = V_{GS0} - V_{DD} + V_{\text{on}} \tag{5.33}$$

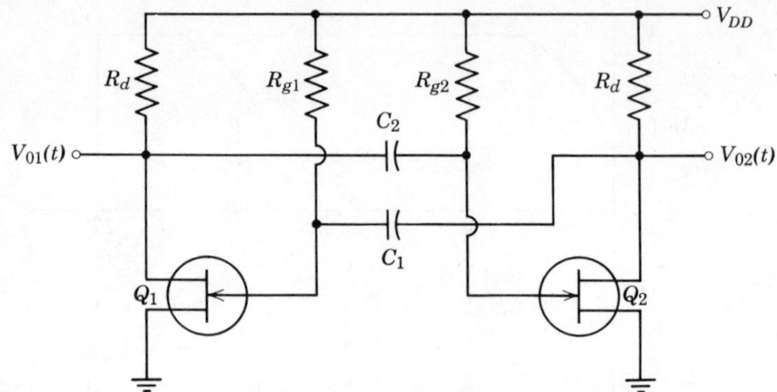

Figure 5.10. Astable JFET circuit.

Now the capacitor C_2 is charged through the resistance R_{g2}, since $V_{D1}(t)= V_{\text{on}}$ during this part of the cycle. Hence the time constant is

$$\tau_s = C_2 R_{g2} = CR_g \tag{5.34}$$

Since $V_{G2}(t)$ will rise to the final value, $V_{G2}(\infty)= V_{DD}$, if undisturbed, we have (see Chapter 1)

$$\begin{aligned} V_{G2}(t) &= V_{G2}(0^+)\exp\left(-\frac{t}{\tau_s}\right) + V_{G2}(\infty)\left[1-\exp\left(-\frac{t}{\tau_s}\right)\right] \\ &= V_{DD} - (2V_{DD} - V_{GS0} - V_{\text{on}})\exp\left(-\frac{t}{\tau_s}\right) \end{aligned} \tag{5.35}$$

The FET Q_2 switches on when $V_{G2}(t)= V_p$. Let this occur at $t= T_s$; then

$$\exp\left(-\frac{T_s}{\tau_s}\right) = \frac{V_{DD} - V_p}{2V_{DD} - V_{GS0} - V_{\text{on}}} \quad \text{or} \quad T_s = CR\ln\left(\frac{2V_{DD} - V_{GS0} - V_{\text{on}}}{V_{DD} - V_p}\right) \tag{5.36}$$

so that the full period T of the waveform is

$$T=2T_s=2CR_g\ln\left(\frac{2V_{DD} - V_{GS0} - V_{\text{on}}}{V_{DD} - V_p}\right) \tag{5.36a}$$

If $|V_{DD}| \gg |V_p|$ and $|V_{DD}| \gg |V_{GS0} - V_{\text{on}}|$, this reduces to

$$T=2T_s \cong 1.38\,CR_g \tag{5.36b}$$

Usually the more accurate expression for T is needed.

When Q_2 switches on, $V_{G2}(t)$ suddenly rises from V_p to V_{GS0}. Therefore $V_{D1}(t)$, which was equal to V_{on} at $t= T_s^-$, becomes equal to $V_{\text{on}} + V_{GS0} -$

V_p at $t=T_s^+$ because the voltage on the capacitor C cannot change abruptly. Now the capacitance C charges through R_d, since $V_{G2}(t)=V_{GS0}$ during this part of the waveform. The time constant involved is thus

$$\tau_R = CR_d \tag{5.37}$$

and the final value of $V_{D1}(t)$ is $V_{D1}(\infty)=V_{DD}$. It takes the time T_R, the recovery time, for $V_{D1}(t)$ to reach this value, where

$$T_R \cong 4\tau_R = 4CR_d \tag{5.38}$$

We now require

$$T_R < T_s \quad \text{or} \quad 4CR_d < CR_g \ln 2 \quad \text{or} \quad R_g > \frac{4}{\ln 2} R_d = 5.8 R_d \tag{5.39}$$

If R_g is too large, however, this slows down the circuit unduly. For example, $R_g = 10R_d$ is usually sufficient.

Figure 5.11 shows the generated waveform for $V_{D1}(t)$ and $V_{G2}(t)$.

Example 2. Find the full period of the waveform if $R_d=5$ KΩ, $R_g=50$ KΩ, $C=1000$ pF, $V_{DD}=20$ V, $V_p=-3$ V, $V_{GS0}=0.70$ V, and $V_{on}=1.0$ V.

Answer.

$$T = 2T_1 = 2CR_g \ln\left(\frac{2V_{DD} - V_{GS0} - V_{on}}{V_{DD} - V_p}\right)$$

$$= 2\times 10^{-9} \times 5 \times 10^4 \ln\left(\frac{38.3}{23.0}\right) = 0.51 \times 10^{-4} \text{ s},$$

so that the approximation $T = 2CR_g \ln 2 = 0.69 \times 10^{-4}$ s is not too accurate in this case.

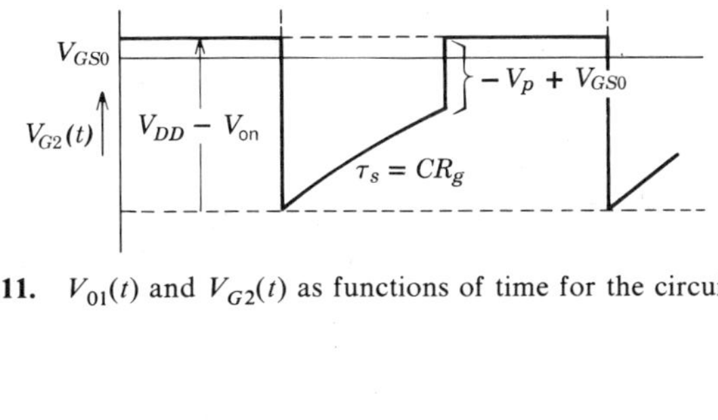

Figure 5.11. $V_{01}(t)$ and $V_{G2}(t)$ as functions of time for the circuit in Fig. 5.10.

5.4 THE DIRECT-COUPLED SCHMITT TRIGGER

A transistorized Schmitt trigger is a flip-flop circuit which is emitter-coupled. There are many versions of it; we discuss the direct-coupled Schmitt trigger shown in Fig. 5.12. There are two possible stable states:

1. Q_1 is off; Q_2 is on.
2. Q_1 is on; Q_2 is off.

The resistors R_1, R_2, and R_3, and the voltages V_A, V_{CC}, V_{EE}, are chosen so that the neither transistor is ever saturated. Switching occurs when the transistor that was off just begins to draw current.

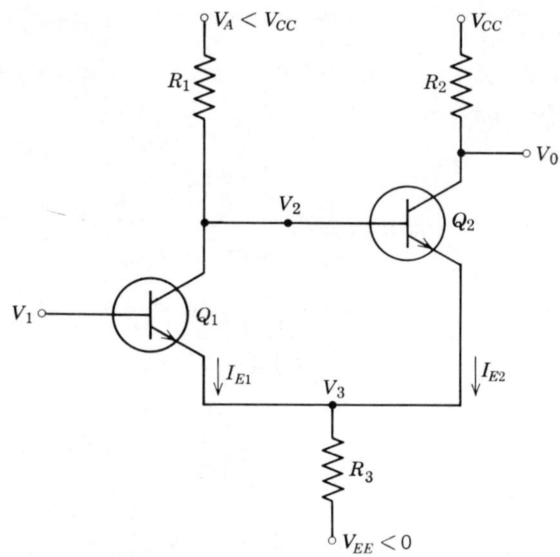

Figure 5.12. Emitter-coupled Schmitt trigger circuit using transistors.

We now show that, when V_0 is plotted versus V_1, a hysteresis loop occurs. There is a high value, $V_{OH} = V_{CC}$, when Q_2 is off, and a low value V_{OL} when Q_1 is off, which needs to be calculated. If V_1 is increased from a low value, so that Q_2 is on, $V_0 = V_{0L}$ for $V_1 \leqslant V_{1H}$. If V_1 is then decreased from a high value, so that Q_2 is off, then $V_0 = V_{0H}$ for $V_1 \geqslant V_{1L}$. V_{1H} and V_{1L} also need to be calculated.

Suppose Q_1 is off; then $V_2 = V_A$ (we neglect the small base current of Q_2). If V_1 is 0.10 V lower than V_2, then Q_1 just begins to draw current.

That is, to have Q_1 off, we require*

$$V_1 \leq (V_A - 0.10) \tag{5.40}$$

Now $V_3 = V_A - V_{BE0}$, where $V_{BE0} \cong 0.70$ V, hence

$$I_{E2} = \frac{V_3 - V_{EE}}{R_3} = \frac{V_A - V_{BE0} - V_{EE}}{R_3} \tag{5.41}$$

and, since $I_{C2} \cong I_{E2}$,

$$V_0 = V_{0L} = V_{CC} - I_{E2}R_2 = V_{CC} - (V_A - V_{BE0} - V_{EE})\frac{R_2}{R_3} \tag{5.42}$$

We now require $V_0 > V_3 + V_{CE,\text{sat}} = V_A - V_{BE0} + V_{CE,\text{sat}}$, to make sure that Q_2 is not saturated.

The value of V_1 for which Q_1 is just starting to draw current is

$$V_1 = V_2 - 0.10 = V_A - 0.10 \tag{5.43}$$

Very close to this value the circuit switches, hence

$$V_{1H} \cong V_A - 0.10 \tag{5.43a}$$

We now calculate V_{1L}. Suppose Q_2 is off; then $V_0 = V_{0H} = V_{CC}$. Moreover,

$$V_3 = V_1 - V_{BE0} = I_{E1}R_3 + V_{EE} \tag{5.44}$$

or

$$I_{E1} = \frac{V_1 - V_{BE0} - V_{EE}}{R_3} \tag{5.44a}$$

Hence, since $I_{C1} \cong I_{E1}$,

$$V_2 = V_A - I_{E1}R_1 = V_A - (V_1 - V_{BE0} - V_{EE})\frac{R_1}{R_3} \tag{5.45}$$

To make sure Q_2 is off, we require

$$V_2 \leq (V_1 - 0.10) \tag{5.46}$$

*In the differential pair formed by Q_1 and Q_2, Q_1 is practically off if $V_1 < V_2 - 0.10$, and Q_2 is practically off if $V_2 < V_1 - 0.10$ (see Section 5.5).

For $V_2 = (V_1 - 0.10)$, Q_2 just begins to draw current; this occurs for a value of V_1 for which

$$V_1 - 0.10 = V_A - (V_1 - V_{BE0} - V_{EE})\frac{R_1}{R_3}$$

Very close to this value of V_1 the circuit switches, hence

$$V_{1L} \cong \frac{V_A + 0.10 + (V_{BE0} + V_{EE})R_1/R_3}{1 + R_1/R_3} \tag{5.47}$$

The hysteresis loop is thus as shown in Fig. 5.13. The circuit has indeed two stable states.

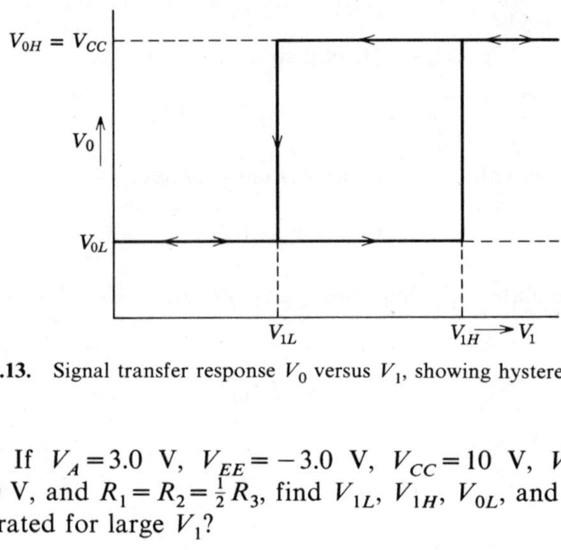

Figure 5.13. Signal transfer response V_0 versus V_1, showing hysteresis effects.

Example 1. If $V_A = 3.0$ V, $V_{EE} = -3.0$ V, $V_{CC} = 10$ V, $V_{BE0} = 0.70$ V, $V_{CE,\text{sat}} = 0.10$ V, and $R_1 = R_2 = \frac{1}{2}R_3$, find V_{1L}, V_{1H}, V_{OL}, and V_{OH}. Can Q_1 become saturated for large V_1?

Answer.

$$V_{OH} = 10.0 \text{ V}; \qquad V_{OL} = 10.0 - (3.0 - 0.7 + 3.0)\tfrac{1}{2} = 7.35 \text{ V}$$

$$V_{1H} = 3.0 - 0.1 = 2.9 \text{ V}; \qquad V_{1L} = \frac{3.0 + 0.1 + (0.7 - 3.0)\frac{1}{2}}{1.5} = 1.30 \text{ V}$$

Q_1 will saturate when V_2 becomes smaller than $V_1 - V_{BE0} + V_{CE,\text{sat}}$, which can occur for large V_1, according to Eq. (5.45). Real difficulties occur when V_1 becomes much larger than this.

5.5 THE DIFFERENTIAL AMPLIFIER

5.5a The Differential Transistor Amplifier

We now prove that a voltage difference of about 0.10 V is needed to turn one half of a differential transistor pair off.

Figure 5.14*a* shows a differential transistor pair with input voltages V_1 and V_2. Here I_0 is a current generator keeping the total current through the pair constant. The voltages V_1 and V_2 are used to switch current from one transistor to the other, and vice versa. The voltage V_E between the emitter

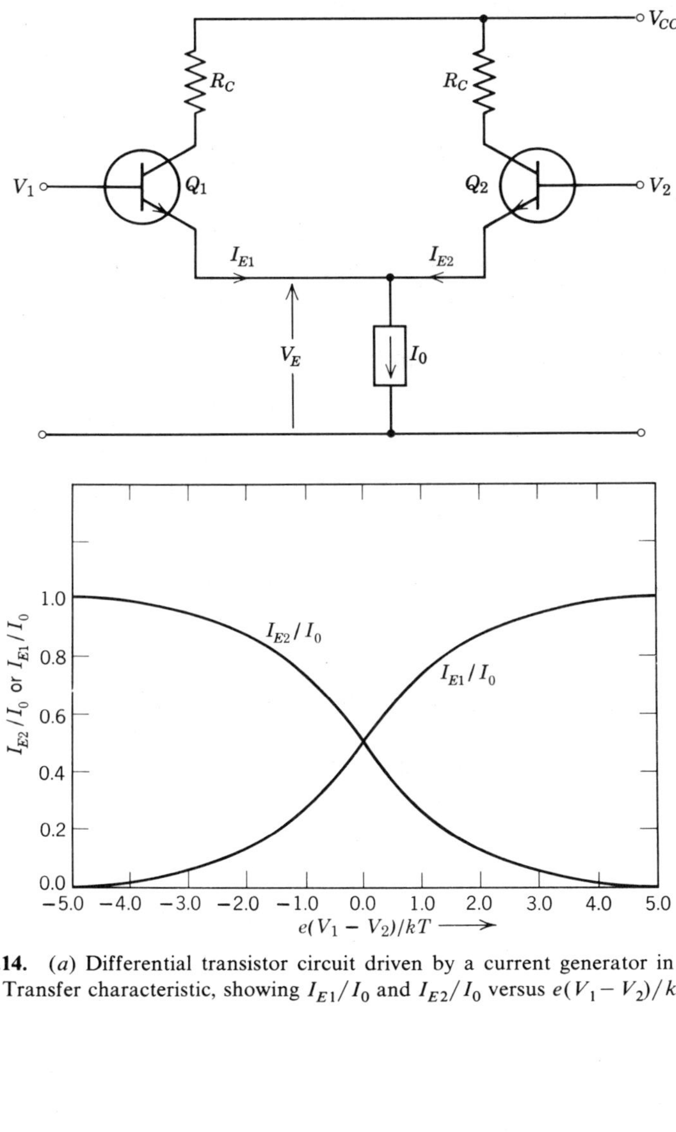

Figure 5.14. (*a*) Differential transistor circuit driven by a current generator in the emitter lead. (*b*) Transfer characteristic, showing I_{E1}/I_0 and I_{E2}/I_0 versus $e(V_1 - V_2)/kT$.

and the ground depends on V_1 and V_2. We have

$$I_{E1}=I_{Es}\exp\left[\frac{e(V_1-V_E)}{kT}\right] \qquad I_{E2}=I_{Es}\exp\left[\frac{e(V_2-V_E)}{kT}\right] \tag{5.48}$$

But $I_{E1}+I_{E2}=I_0$. Substituting Eq. (5.48) and solving for $I_{Es}\exp(-eV_E/kT)$ yields

$$I_{Es}\exp\left(\frac{-eV_E}{kT}\right)=\frac{I_0}{\exp(eV_1/kT)+\exp(eV_2/kT)} \tag{5.49}$$

Hence, substituting back into Eq. (5.48),

$$I_{E1}=\frac{I_0}{1+\exp[e(V_2-V_1)/kT]} \qquad I_{E2}=\frac{I_0}{1+\exp[e(V_1-V_2)/kT]} \tag{5.50}$$

Therefore $I_{E2}\cong 0.01I_0$ if $e(V_1-V_2)/kT\cong 4.60$ or $V_1-V_2\cong 0.12$ V, so current just begins to flow through Q_2 if $V_1-V_2\cong 0.10$ V, as required. Figure 5.14*b* shows I_{E1}/I_0 and I_{E2}/I_0 as functions of $e(V_1-V_2)/kT$.

5.5b The Differential FET Amplifier

Figure 5.15*a* shows a differential FET pair with input voltages V_1 and V_2; I_0 is a current generator keeping the total current through the pair constant. The voltage V_s between the source and the ground depends on V_1 and V_2. We have, if $K_1=K_2=K$,

$$I_{D1}=K(V_1-V_s-V_p)^2 \qquad I_{D2}=K(V_2-V_s-V_p)^2 \tag{5.51}$$

$$I_0=I_{D1}+I_{D2} \tag{5.51a}$$

In contrast with the transistor case, the solution cannot be put in a simple form.

It is easily seen that $I_{D2}=0$ if $V_2-V_s-V_p=0$, in which case $I_{D1}=I_0$ and $V_1-V_s-V_p=(I_0/K)^{1/2}$, or $V_1-V_2=(I_0/K)^{1/2}$. In the same way, $I_{D1}=0$ if $V_1-V_s-V_p=0$, $I_{D2}=I_0$, and $V_2-V_s-V_p=(I_0/K)^{1/2}$, or $V_1-V_2=-(I_0/K)^{1/2}$. For

$$-\left(\frac{I_0}{K}\right)^{1/2}\leqslant V_1-V_2\leqslant\left(\frac{I_0}{K}\right)^{1/2} \tag{5.51b}$$

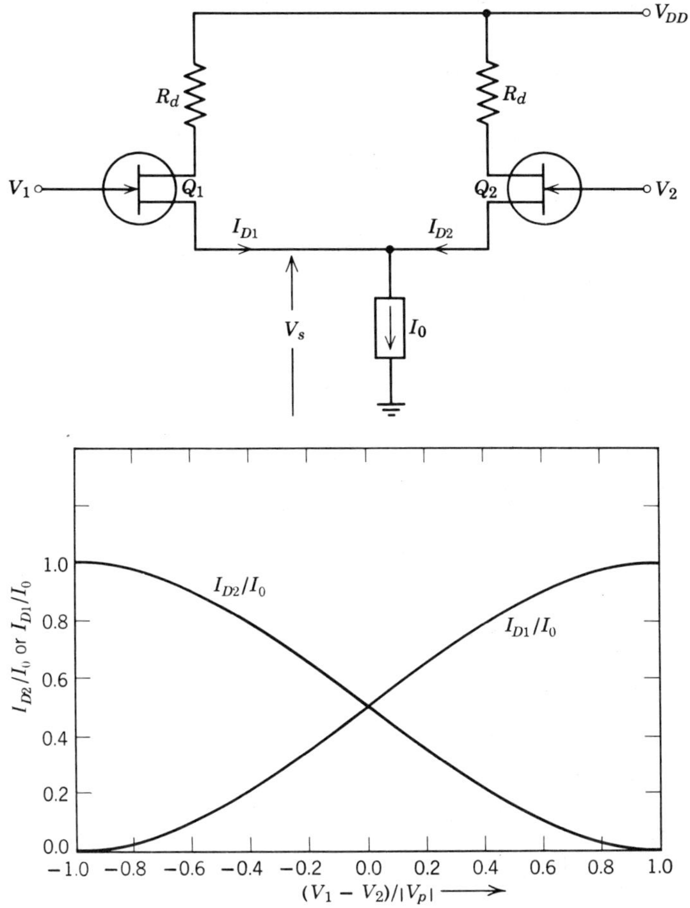

Figure 5.15. (*a*) Differential JFET circuit driven by a current generator in the common source lead. (*b*) Transfer characteristic showing I_{D1}/I_0 and I_{D2}/I_0 versus $(V_1 - V_2)/|V_p|$.

the current is thus switched from one transistor to the other.

Figure 5.15*b* shows I_{D1}/I_0 and I_{D2}/I_0 as functions of $(V_1 - V_2)/|V_p|$ for $I_0 = 4$ mA, $K = 1$ mA/V^2, $V_p = -2.0$ V, and $V_2 = 6.0$ V. This result is best calculated numerically.

PROBLEMS

5.1 If Q_1 is off and triggered on at $t = 0$, find the output waveform for the circuit in Fig. 5.1*a* if $V_{CC} = 4$ V, $R_1 = 1000\ \Omega$, $R_c = 500\ \Omega$, $V_{BE0} = 0.70$ V, $V_{CE,\text{sat}} = 0.10$ V, and $\beta_F = 50$..

5.2 If Q_1 is off and triggered on at $t=0$, find the output waveform in Fig. 5.2*a* if $V_{DD}=30$ V, $V_{GG}=-30$ V, $V_p=-3$ V, $R_{\text{on}}=300\ \Omega$, $R_d=5000\ \Omega$, $R_1=10{,}000\ \Omega$, and $R_2=15{,}000\ \Omega$.

5.3 Repeat Problem 5.2 for the circuit in Fig. 5.3. $V_{DD}=-20$ V, $V_T=-4$ V, and $K_1/K_2=4.0$. Find the accurate values of V_{02} in both cases.

5.4 In the circuit in Fig. 5.1*b*, $\tau_T=5\times10^{-8}$ s. Find the capacitance C_1 needed to speed up the turn-on transient if $R_1=2000\ \Omega$. If $R_C=500\ \Omega$, find the net speedup in the response.

5.5 In the circuit in Fig. 5.2*b*, $R_d=5000\ \Omega$, $R_1=R_2=10{,}000\ \Omega$, and $C_{gs}=5$ pF.

(a) Find C_1 so that $C_1R_1=C_{gs}R_2$.

(b) Find the net speedup in the response.

5.6 In the circuit in Fig. 5.4, $R=10{,}000\ \Omega$, $R_1=2000\ \Omega$, $R_c=1000\ \Omega$, $V_{CC}=4.0$ V, $V_{BE0}=0.70$ V, $V_{CE,\text{sat}}\cong0.00$ V, and Q_1 is triggered at $t=0$.

(a) Choose the capacitance C so that the duration T_s of the output pulse is 1.0 μs. Use the accurate expression for T_s.
(b) Find the accurate height of the output pulse if it is taken from the collector of Q_2.
(c) Repeat (b) if the output is taken from the collector of Q_1.

5.7 Repeat Problem 5.6 for $V_{CC}=6.0$ V, $BV_{BE0}=3.5$ V, $V_{BE0}=0.70$ V, $V_{CE,\text{sat}}=0.00$ V, $\beta_F=50$, $R=10{,}000\ \Omega$, $R_1=2000\ \Omega$, and $R_c=1000\ \Omega$. Here $-BV_{BE0}$ is the voltage at which the base-emitter junction breaks down.

5.8 In the circuit in Fig. 5.6, $V_{DD}=-20$ V, $V_T=-4.0$ V, $R=10{,}000\ \Omega$, $R_d=5000\ \Omega$, and $V_{\text{on}}=-1.0$ V.

(a) Find the capacitance C so that the length of the pulse is 10^{-6} s.
(b) Find the height of the output pulse if it is taken from the drain of Q_2.
(c) Repeat (b) if the output pulse is taken from the drain of Q_1.

5.9 In the circuit in Fig. 5.8, $V_{CC}=4.0$ V, $V_{BE}=0.70$ V, $V_{CE,\text{sat}}=0.00$ V, $\beta_F=50$, $R_c=1000\ \Omega$, $R_1=R_2=R=10{,}000\ \Omega$, $C=C_1=C_2$ is the coupling capacitance. Find the capacitance C so that the period of the generated square wave is 10^{-4} s. Use the accurate expression for T_{s1}.

5.10 Find R and C for the monostable circuit given in Fig. 5.16 if the output is to be a positive pulse of 10-μs duration following transition to the quasi-stable state. Make sure R is not too large, or Q_2 will not be saturated at a steady state. Sketch and label waveshapes at both the bases and the collectors; $Q_1=Q_2$, $\beta_F=40$, $V_{BE0}\cong0.70$ V, and $V_{CE,\text{sat}}\cong0.20$ V.

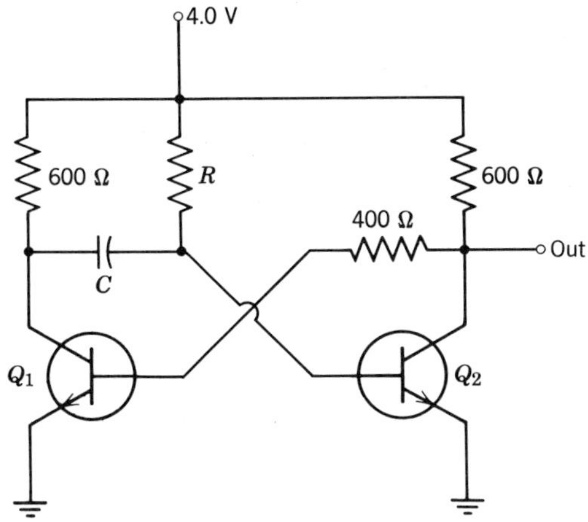

Figure 5.16

5.11 In the circuit in Fig. 5.10, $V_{DD}=20$ V, $V_p=-3.0$ V, $V_{GS0}=0.70$ V, $V_{on}=1.0$ V, $R_d=5$ KΩ, and $R_g=50$ KΩ. Find the capacitance C needed to make a square wave with a period of 10^{-4} s.

5.12 The transistors of the circuit in Fig. 5.12 have $V_{BE0}=0.70$ V and $V_{CE,\text{sat}}=0.10$ V. If $V_{CC}=10$ V, $V_{EE}=-4.0$ V, $V_A=4.0$ V, $R_3=2000$ Ω, and $R_1=R_2=1000$ Ω, find the values of V_{1L}, V_{1H}, V_{0L}, and V_{0H}. Determine whether or not all the transistors are nonsaturated at all times.

CHAPTER

6 Boolean Algebra and Its Implementation

Section 6.1 discusses the rules of Boolean algebra and its realizations. Section 6.2 deals with wired logic as a means of obtaining new logic functions by connecting the outputs of two logic circuits together. Section 6.3 applies the rules of Boolean algebra to binary arithmetic and discusses ways of realizing this arithmetic.

6.1 BOOLEAN ALGEBRA AND ITS REALIZATIONS

We state here the rules of Boolean algebra in some detail and then investigate how the various relationships can be implemented electronically. See also Section 2.2c.

6.1a Addition: $A+B$

1. $A+B+C=(A+B)+C=A+(B+C)$
2. $A+B=B+A$
3. $A+A=A$
4. $A+1=1$
5. $A+0=A$

The third relation can be proved with the help of a truth table.

A	$A+A$
0	0
1	1

Realization: the or circuit in Fig. 2.10*a*.

6.1b Multiplication: $A \cdot B$

1. $ABC = (AB)C = A(BC)$
2. $AB = BA$
3. $A \cdot A = A$
4. $A \cdot 1 = A$
5. $A \cdot 0 = 0$

The third relation can be proved with the help of a truth table.

A	$A \cdot A$
0	0
1	1

Realization: the and circuit in Fig. 2.10*b*.

6.1c Mixed Relations

1. $A(B+C) = AB + AC$
2. $A + AB = A$
3. $(A+B)(A+C) = A + BC$

Proof of the second relation:

$$A + AB = A(1+B) = A \cdot 1 = A$$

Proof of the third relation:

$$\begin{aligned}(A+B)(A+C) &= A \cdot A + AB + AC + BC \\ &= A + AB + AC + BC \\ &= A(1+B+C) + BC = A + BC\end{aligned}$$

since $1 + B + C = 1$

6.1d **Inversion** (Not Operation)

1. $\overline{\overline{A}}=A$
2. $A+\overline{A}=1$
3. $A\cdot\overline{A}=0$
4. $A+\overline{A}B=A+B$
5. $\overline{A}+AB=\overline{A}+B$

Proof of the fourth relation:

$$A+\overline{A}B=A(1+B)+\overline{A}B=A+(A+\overline{A})B=A+B$$

Proof with the help of a truth table:

A	B	$\overline{A}$	$\overline{A}B$	$A+\overline{A}B$	$A+B$
0	0	1	0	0	0
1	0	0	0	1	1
0	1	1	1	1	1
1	1	0	0	1	1

Realization: the transistor inverter circuit and the MOSFET inverter circuit.

6.1e **DeMorgan's Laws**

1. $\overline{A+B+C+\cdots}=\overline{A}\,\overline{B}\,\overline{C}\cdots$
2. $\overline{ABC\cdots}=\overline{A}+\overline{B}+\overline{C}+\cdots$

Proof of the first law: If at least one of the symbols, for example, C is a 1, then $A+B+C+\cdots=1$ and $\overline{A+B+C+\cdots}=0$. But in this case $\overline{C}=0$, hence $\overline{A}\,\overline{B}\,\overline{C}\cdots=0$. If *all* the symbols are 0's, all the inverse symbols are 1's. Hence $A+B+C+\cdots=0$, $\overline{A+B+C+\cdots}=1$, and $\overline{A}\,\overline{B}\,\overline{C}\cdots=1$.

Proof of the second law: If at least one of the symbols, for example, C, is a 0, then $ABC\cdots=0$, hence $\overline{ABC\cdots}=1$. But then $\overline{C}=1$, hence $\overline{A}+\overline{B}+\overline{C}+\cdots=1$. If all the symbols are 1's, then $ABC\cdots=1$, hence $\overline{ABC\cdots}=0$. But since all the inverses are 0's, $\overline{A}+\overline{B}+\overline{C}+\cdots=0$.

Realization of $\overline{A+B+C+\cdots}$: an or circuit followed by an inverter. This is called a *nor* circuit.

Realization of $\overline{ABC\cdots}$: an and circuit followed by an inverter. This is called a *nand* circuit.

6.1f Symbols

Symbols for the various logic circuits, also called gates, are shown in Fig. 6.1. Figure 6.1*a* represents an or circuit, Fig. 6.1*c* an and circuit, and Fig. 6.1*e* an inverter circuit.

An inverter circuit is here represented by a circle. For example, Fig. 6.1*b* represents an or circuit followed by an inverter; it thus is a nor circuit. Figure 6.1*d* is an and circuit followed by an inverter; it thus represents a nand circuit. When the inputs of a nor or a nand gate are connected, one obtains an inverter circuit (compare Fig. 6.1*f* and *g*).

One can also put an inverter between the input terminal and the gate, as indicated by a circle at the input. Examples are given in Figs. 6.2*a–d*.

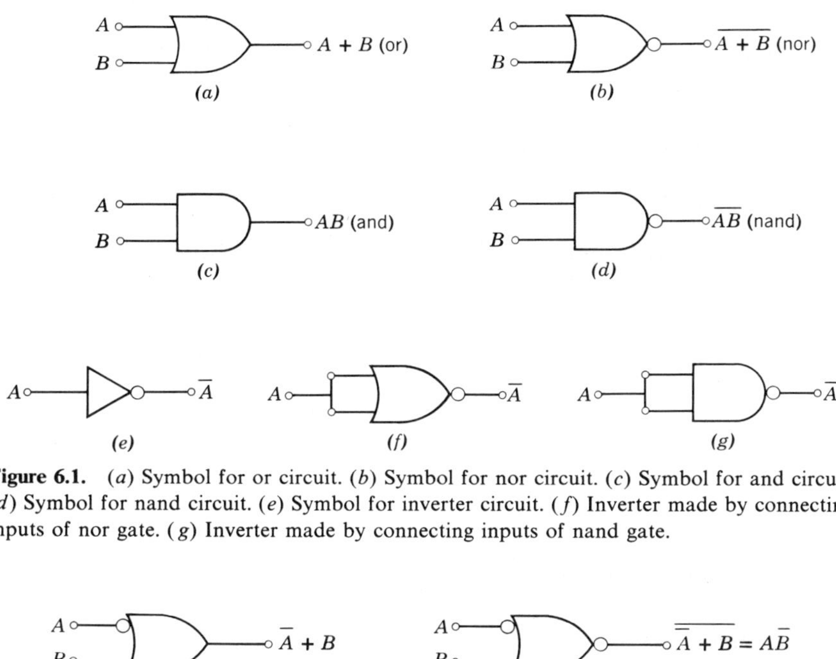

Figure 6.1. (*a*) Symbol for or circuit. (*b*) Symbol for nor circuit. (*c*) Symbol for and circuit. (*d*) Symbol for nand circuit. (*e*) Symbol for inverter circuit. (*f*) Inverter made by connecting inputs of nor gate. (*g*) Inverter made by connecting inputs of nand gate.

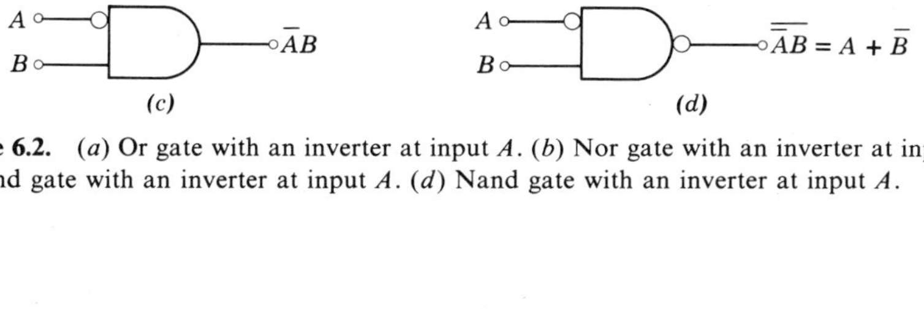

Figure 6.2. (*a*) Or gate with an inverter at input *A*. (*b*) Nor gate with an inverter at input *A*. (*c*) And gate with an inverter at input *A*. (*d*) Nand gate with an inverter at input *A*.

Figure 6.2*b* shows the circuit $\overline{\bar{A}+B}=\bar{\bar{A}}\bar{B}=A\bar{B}$. Figure 6.2*d* shows the circuit $\overline{\bar{A}B}=\bar{\bar{A}}+\bar{B}=A+\bar{B}$.

6.1g Inhibit and Enable Operations

If in an and circuit one of the input terminals, designated by S, feeds into the circuit through an inverter, then we have (Fig. 6.3*a*):

1. For $S=0$, $Y=ABC$. This is called the *enable* operation, since the S-terminal now *enables* the and gate to function.

2. For $S=1$, $Y=0$. This is called the *inhibit* operation, since the S-terminal inhibits the operation of the and gate.

If in an or circuit one of the input terminals S feeds into the circuit through an inverter, then we have (Fig. 6.3b):

1. For $S=1$, $Y=A+B+C$. *Enable* operation.

2. For $S=0$, $Y=1$. *Inhibit* operation.

Again, the enable operation permits the or circuit to function.

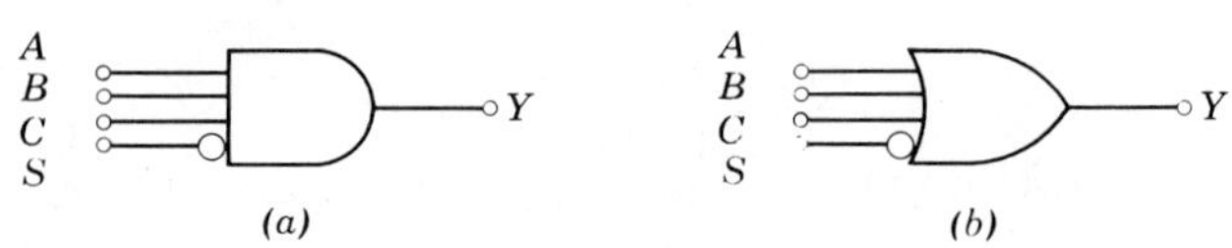

Figure 6.3. Enable and inhibit circuits.(*a*) Four-input and gate with an inverter at the input S. (*b*) Four-input or gate with an inverter at the input S.

6.1h Exclusive Or: $Y=(A+B)(\overline{AB})=A\oplus B$

This means either A or B but not both.

A	B	$A+B$	$\overline{AB}$	Y
0	0	0	1	0
1	0	1	1	1
0	1	1	1	1
1	1	1	0	0

That is, $Y=0$ for $A=B$ and $Y=1$ for $A\neq B$. This definition is realized in

Fig. 6.4*a*. Other forms of the exclusive or are:

1. $Y=(A+B)(\overline{AB})=(\bar{A}+\bar{B})(A+B)$. This definition is realized in Fig. 6.4*b*.

2. $Y=(A+B)(\bar{A}+\bar{B})=A\bar{A}+\bar{A}B+A\bar{B}+B\bar{B}=\bar{A}B+A\bar{B}$. This relationship is realized in Fig. 6.4*c*.

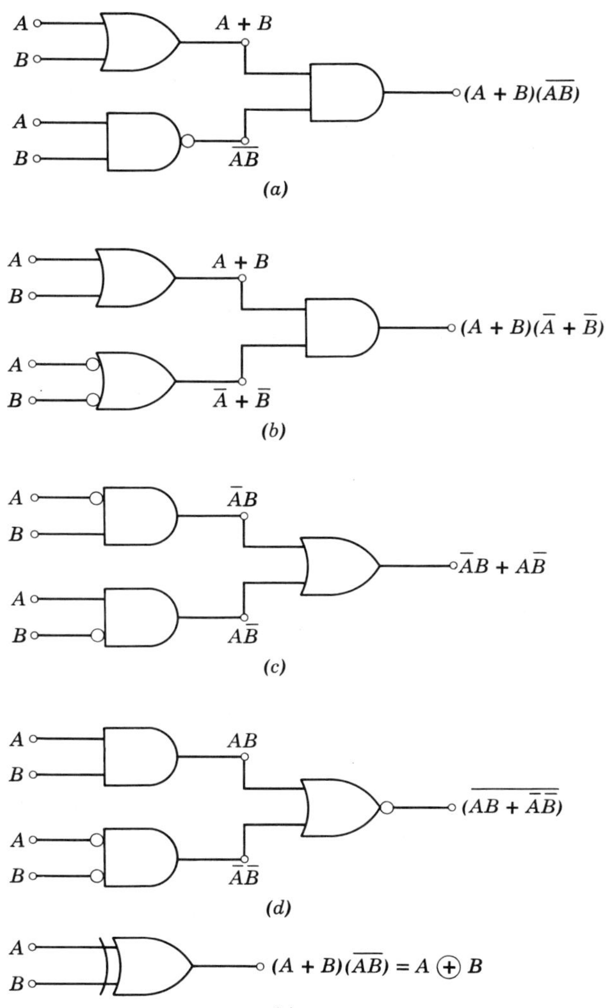

Figure 6.4. Various realizations of the exclusive or $A\oplus B$. (*a*) $A\oplus B=(A+B)(\overline{AB})$. (*b*) $A\oplus B=(A+B)(\bar{A}+\bar{B})$. (*c*) $A\oplus B=\bar{A}B+A\bar{B}$. (*d*) $A\oplus B=\overline{(AB+\bar{A}\bar{B})}$. (*e*) Symbol for $A\oplus B$.

3. $Y=\overline{(AB+\bar{A}\bar{B})}$ for the exclusive or. This relationship is realized in Fig. 6.4*d*.

Proof: The exclusive or means $Y=0$ for $A=B$, and $Y=1$ for $A\neq B$. $A=B$: $Y=\overline{(AA+\bar{A}\bar{A})}=\overline{(A+\bar{A})}=\bar{1}=0$, which fits. $A\neq B$: $AB=\bar{A}\bar{B}=0$, hence $\overline{(AB+\bar{A}\bar{B})}=1$, which also fits.

Alternate proof:

$$Y=\left(\overline{AB+\bar{A}\bar{B}}\right)=(\overline{AB})\left(\overline{\bar{A}\bar{B}}\right)=(\bar{A}+\bar{B})\left(\bar{\bar{A}}+\bar{\bar{B}}\right)=(\bar{A}+\bar{B})(A+B)$$

A shorthand symbol for the exclusive or is pictured in Fig. 6.4*e*.

6.1i Equivalence of different circuits

Not all circuits or, and, nor, and nand are necessary. Figure 6.5*a–d* shows nand, and, nor, and or circuits, respectively. They are made by using inverters at the input terminals; the and and or circuits also have an inverter at the output terminal.

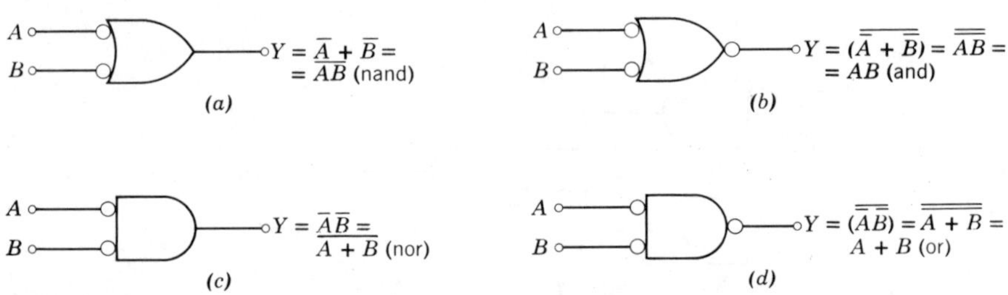

Figure 6.5. Various forms of nand, and, nor, or gates. (*a*) Or gate with two inverters at inputs. (*b*) Nor gate with two inverters at inputs. (*c*) And gate with two inverters at inputs. (*d*) Nand gate with two inverters at inputs.

6.1j Positive Logic and Negative Logic

1. An and circuit for positive logic functions is an or circuit for negative logic functions.

2. An or circuit for positive logic functions is an and circuit for negative logic functions.

An and circuit for positive logic is symbolized as

$$Y=ABC\cdots$$

Changeover from positive logic to negative logic means that all 0's become 1's and all 1's become 0's. Hence Y becomes $\overline{Y}$, A becomes $\overline{A}$, B becomes $\overline{B}$, and so on.

$$\overline{Y} = \overline{ABC\cdots} = \overline{A} + \overline{B} + \overline{C} + \cdots$$

which is a *negative or*.

An or circuit for positive logic has

$$Y = A + B + C + \cdots$$

But Y becomes $\overline{Y}$, A becomes $\overline{A}$, B becomes $\overline{B}$, and so on, when changing over to negative logic. Hence

$$\overline{Y} = \overline{A + B + C + \cdots} = \overline{A}\,\overline{B}\,\overline{C}\cdots$$

which is a *negative and*.

We noticed these relationships earlier in Chapter 2.

6.2 WIRED LOGIC

By wiring the outputs of several gates together, one obtains new logic operations without additional circuitry. This is called *wired logic*.

The basis of wired logic is that the output of an inverter, nor, or nand circuit is a short-circuit if the output is a 0 and an open-circuit if the output is a 1. Hence the output of the combination is a 0 unless both outputs give a 1 simultaneously. Therefore the output is the *product* of the individual outputs $Y = Y_1 Y_2$.

We now give several examples:

1. *Nand gate and inverter connected at output.*

This connection is pictured in Fig. 6.6*a*. It can be described as

$$Y = (\overline{AB})\overline{C}$$

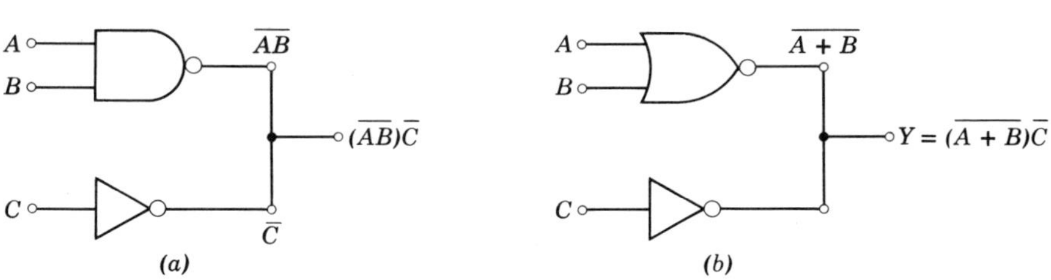

Figure 6.6. Examples of wierd logic using nand and nor gates. (*a*) Nand circuit and inverter wired together, giving rise to the logic operation $(\overline{AB+C})$, corresponding to an enable circuit. (*b*) Nor circuit and inverter wired together, giving an extended nor gate, corresponding to an enable circuit.

which corresponds to an inhibit-enable circuit:

$$Y=0 \qquad \text{if } \overline{C}=0 \text{ or } C=1\text{: inhibit operation}$$

$$Y=\overline{AB} \qquad \text{if } \overline{C}=1 \text{ or } C=0\text{: enable operation}$$

2. *Nor gate and inverter connected at output.*

This connection is pictured in Fig. 6.6*b*. It can be described as

$$Y=(\overline{A+B})\overline{C}$$

which corresponds to an inhibit-enable circuit:

$$Y=0 \qquad \text{if } \overline{C}=0 \text{ or } C=1\text{: inhibit operation}$$

$$Y=\overline{A+B} \qquad \text{if } \overline{C}=1 \text{ or } C=0\text{: enable operation}$$

3. *Two inverters wired together.*

This connection is pictured in Fig. 6.7*a* and can be described as

$$Y=\overline{A}\,\overline{B}=\overline{A+B}$$

Hence this is a *nor* circuit.

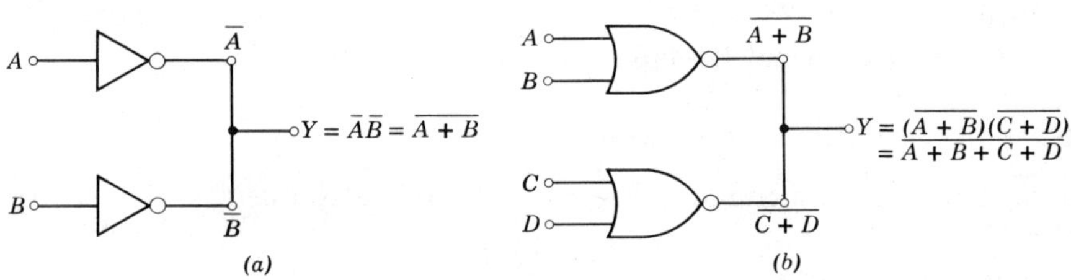

Figure 6.7. Other examples of wired logic. (*a*) Nor gate made of two inverters wired together. (*b*) Extended nor gate made of two nor gates wired together.

4. *Two nor gates wired together.*

This connection is pictured in Fig. 6.7*b* and can be described as

$$Y=(\overline{A+B})(\overline{C+D})=\overline{A+B+C+D}$$

This is just as if there were four inputs at one gate; it is nothing but a four-input *nor* circuit.

5. *Nor and nand gates wired together.*

This connection is shown in Fig. 6.8*a* and can be described as

$$Y=(\overline{A+B})(\overline{CD})=\overline{A+B+CD}$$

This is a new logic operation.

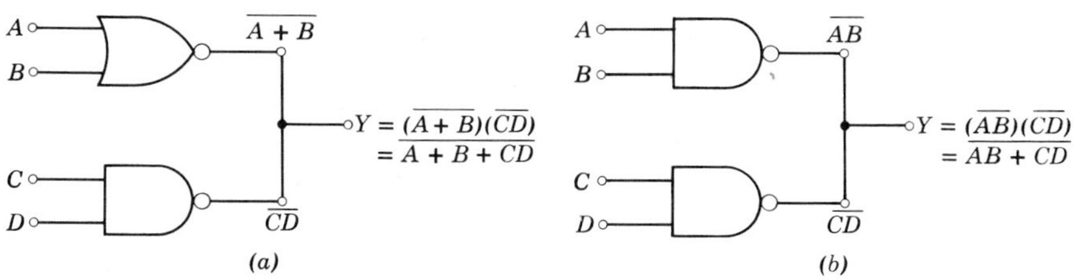

Figure 6.8. New logic operations involving wired logic.(*a*) Nor and nand gate wired together $[Y=(\overline{A+B+CD})]$. (*b*) Two nand gates wired together $[Y=(\overline{AB+CD})]$.

6. *Two nand gates wired together*

This connection is shown in Fig. 6.8*b* and can be described as

$$Y=(\overline{AB})(\overline{CD})=\overline{AB+CD}$$

This is also a new logic operation.

In each of the cases discussed, the output Y of the wired gates is Y_1Y_2, where Y_1 and Y_2 are the outputs of the individual gates. For this reason these gates are called *wired and* gates or *implied and* gates. For each individual gate a 0 output is developed across a short circuit and a 1 output is developed across an open circuit. This situation does not hold for emitter-coupled logic (ELC), and therefore this form of wired logic requires a separate discussion. We come back to it in Chapter 7.

6.3 BINARY ADDITION

First, we write down a general rule for expressing a decimal number in binary form: Write down the number to the right hand of the line. Divide by 2, write the remainder under the number and the resulting quotient to the left of it. Repeat this until you cannot go any further.

Example 1. Write the number 135 in binary form.

Answer.

1	2	4	8	16	33	67	135
1	0	0	0	0	1	1	1

Therefore the binary equivalent is 10000111.

The laws of binary addition are the following:

$$\begin{array}{ll} 0+0=0 & \text{carry } 0 \\ 1+0=1 & \text{carry } 0 \\ 0+1=1 & \text{carry } 0 \\ 1+1=0 & \text{carry } 1 \end{array}$$

The first column is called the sum S, and the second column is called the "carry" signal C that must be added to the next set of digits.

Example 2. Add 111 and 1001.

Answer. $111+1001=10000$.

We can write S and C as

$$S\,(\text{sum}) = X \oplus Y \text{ (exclusive or)}$$

$$C\,(\text{carry}) = XY$$

The combination is called a *half adder*. It is pictured in Fig. 6.9.

By adding the carry signal from the previous operation we obtain a full adder. A *full adder* consists of two half adders and an or circuit. It is pictured in Fig. 6.10.

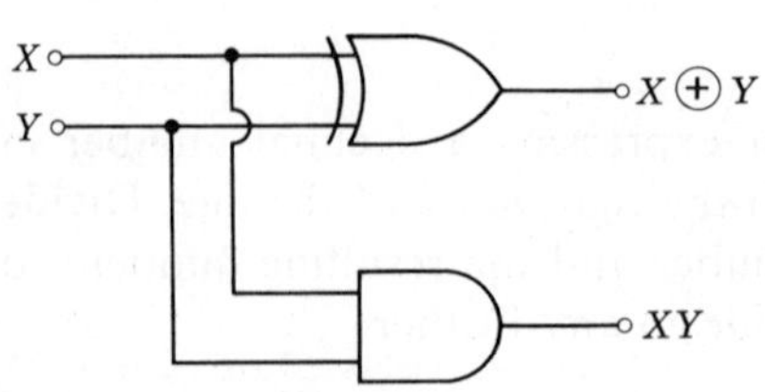

Figure 6.9. Half adder.

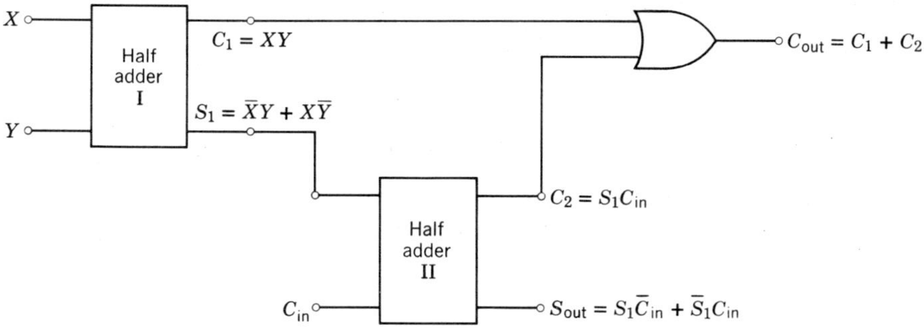

Figure 6.10. Full adder consisting of two half adders and an or circuit.

We now express C_{out} and S_{out} in Boolean form:

$$C_{out} = C_1 + C_2 = XY + (\bar{X}Y + X\bar{Y})C_{in}$$

$$S_{out} = (\bar{X}Y + X\bar{Y})\bar{C}_{in} + \left(\overline{\bar{X}Y + X\bar{Y}}\right)C_{in}$$

Now

$$\left(\overline{\bar{X}Y + X\bar{Y}}\right) = \left(\overline{\bar{X}Y}\right)\left(\overline{X\bar{Y}}\right) = (X + \bar{Y})(\bar{X} + Y)$$

$$= X\bar{X} + Y\bar{Y} + XY + \bar{X}\bar{Y} = XY + \bar{X}\bar{Y}$$

Hence

$$S_{out} = C_{in}(XY + \bar{X}\bar{Y}) + \bar{C}_{in}(\bar{X}Y + \bar{Y}X)$$

The expression for C_{out} can be further simplified when necessary.

By combining large numbers of full adders in the appropriate manner one can add two arbitrary binary numbers.

PROBLEMS

6.1 Prove that $AB + \bar{A}B + A\bar{B} + \bar{A}\bar{B} = 1$.

6.2 Prove that $ABC + AB\bar{C} + A\bar{B}C + A\bar{B}\bar{C} + \bar{A}BC = A + BC$.

6.3 Prove that $abcd + abc\bar{d} + \bar{a}bc\bar{d} + \bar{a}b\bar{c}\bar{d} + \bar{a}\bar{b}c\bar{d} + \bar{a}\bar{b}\bar{c}\bar{d} = abc + \bar{a}\bar{d}$.

6.4 Show that $a(ab+c)+\bar{a}(\bar{c}+b)=b+ac+\bar{a}\bar{c}$.

6.5 Show that $(\overline{a+b})c+\overline{ac}+ab=\bar{a}+\bar{c}+b$.

6.6 Show that $(A \oplus B)C+AB=(A+B)C+AB$.

6.7 Prove that $\bar{a}+ab=\bar{a}+b$ with the help of a truth table.

6.8 Write down the truth table for $ab+\bar{a}\bar{b}$ and show that it corresponds to $\overline{a \oplus b}$. Can you also prove this from the definition of the exclusive or?

CHAPTER

7 Logic Circuits

In this chapter we discuss a variety of logic circuits. Section 7.1 defines noise immunity, fan-out, and fan-in; it also explains the Schottky barrier diode protection circuit and considers transistor and MOSFET nor and nand gates. Section 7.2 deals with resistor transistor logic (RTL), and Section 7.3 continues with direct coupled transistor logic (DCTL). Section 7.4 explains diode transistor logic (DTL), while Section 7.5 deals with the popular transistor-transistor logic (TTL) gate. Section 7.6 considers various totem-pole circuits which combine a large current capability with short switching times; the limitations set by the collector series resistance are briefly dealt with. Section 7.7 describes emitter-coupled transistor logic circuits (ECL) which have the important feature of fast switching. Section 7.8 discusses MOSFET driver circuits with short switching times, and Section 7.9 summarizes the results of the various logic gates in a table.

7.1 GENERAL CONSIDERATIONS

Before discussing individual logic stages we must first introduce some definitions which are used to judge the quality of logic stages (Section 7.1a), a circuit commonly used to prevent transistors from saturating (Section 7.1b), and nor and nand gates (Section 7.1c).

7.1a Noise Immunity, Fan-out, and Fan-in

When the output of a transistor logic circuit is driven to saturation, the output voltage is equal to $V_{CE,\text{sat}}=0.10-0.20$ V. When the base voltage V_{BE} of a transistor logic circuit is less than $V_{BE1}=0.55$ V, the transistor is off; when the base voltage V_{BE} is equal to $V_{BE0}=0.70$ V, the transistor is fully on.

In order to define the noise immunity, we must distinguish between the case of a 0 input and the case of a 1 input. The noise immunity for a 0 input is defined as

(Maximum input voltage of logic stage for a logical 1 output)

−(output voltage of previous stage for a normal logical 0 output)

The noise immunity for a 1 input is defined as

(Normal input voltage of logic state when on)

−(minimum output voltage of previous stage for which logic state is on)

In the simplest case the noise immunity for a 0 input is $0.70-0.20=0.50$ V, and the noise immunity for a 1 input is $V_{CC}-V_{1H}$ (Fig. 7.1*a*). In other cases the situation can be more complicated. Figure 7.1*a* illustrates the definitions for the case of a transistor with a resistance in series with the base.

Sometimes one wants the noise immunity to be large. This can be achieved by increasing the turn-on voltage of the logic stage, for example, by connecting several diodes in series with its base. Since each diode increases the turn-on voltage of the stage by about 0.70 V, n extra in series increase it by $n\times0.70$ V; consequently, the noise immunity of the system for a 0 input is increased by $0.70n$ V. The noise immunity for a 1 output is decreased, because V_{1H} will increase when V_{1L} is increased; consequently $V_{CC}-V_{1H}$ will decrease.

Often one wants the output of a logic state to drive the inputs of several other logic stages. The number of stages that can be *safely* driven from the output of a giveen logic stage is called the *fan-out* of the stage. Obviously one wants that number to be large. Figure 7.1*b* shows the fan-out of an inverter stage. See also Appendix 4.

Often one wants the input of a logic stage to be driven by the outputs of several previous logic stages. The number of output stages that can *safely* drive the input of a given logic stage is called the *fan-in* of the stage. Obviously one wants this number to be large. Figure 7.1*c* shows the fan-in of an inverter stage.

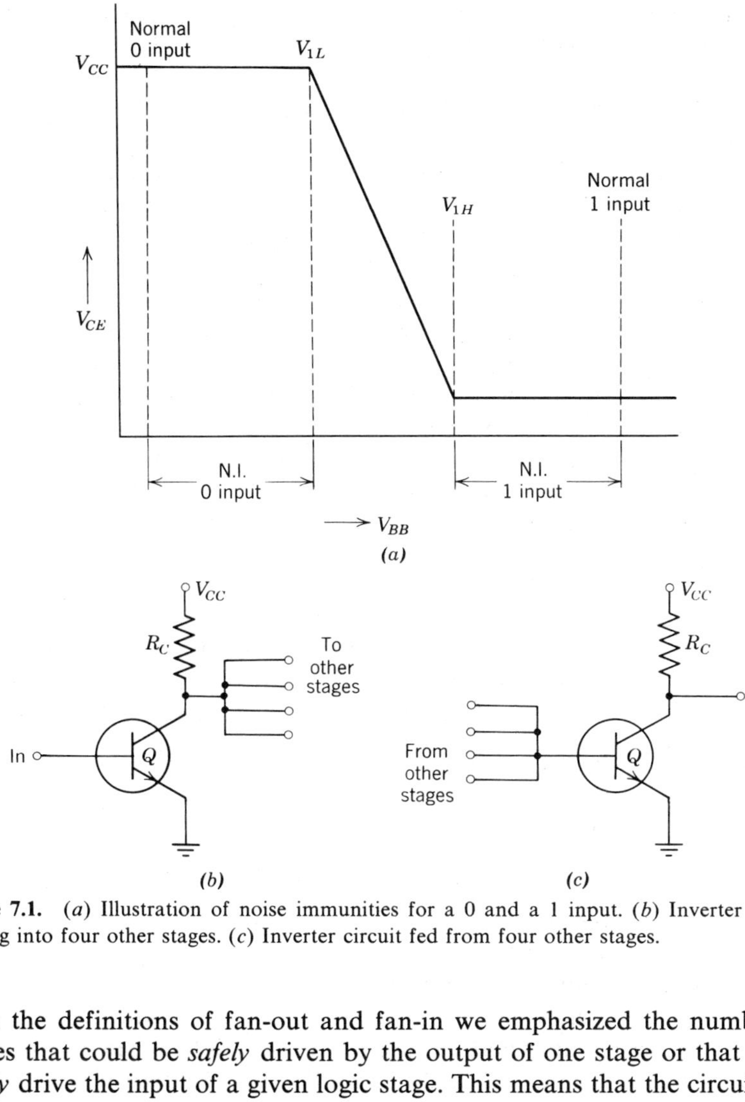

Figure 7.1. (*a*) Illustration of noise immunities for a 0 and a 1 input. (*b*) Inverter circuit feeding into four other stages. (*c*) Inverter circuit fed from four other stages.

In the definitions of fan-out and fan-in we emphasized the number of stages that could be *safely* driven by the output of one stage or that could *safely* drive the input of a given logic stage. This means that the circuits are required to operate even under the worst conditions. Designing fan-out and fan-in circuits thus involves incorporating *worst-case design* criteria. See examples of individual logic gate structures and Appendix 4.

Metal oxide semiconductor (MOS) circuits have a distinct advantage in that the device does not draw any gate current. Theoretically therefore a large number of logic gates can be fed from a single circuit. In practice, however, the number is limited. The chief limitation is that a large fan-out produces a large capacitance in parallel with the output of the logic gate under consideration, and this reduces the switching speed. In practice, compromises must therefore be made.

7.1b Schottky Barrier Diode Protector Circuit

We saw in Chapter 3 that a transistor can be driven deeply into saturation, and that this results in a significant turn-off delay. To prevent this from happening the base-collector diode is often protected by a Schottky barrier diode which clamps the collector-emitter voltage V_{CE} to a value larger than $V_{CE,\text{sat}}$ and so prevents the transistor from being driven to saturation.

To understand how this circuit works, we must first understand the Schottky barrier diode. It is made by evaporating a suitable metal layer onto an n-type semiconductor and making ohmic contacts with the metal (contact A) and with the semiconductor (contact B) (Fig. 7.2a). A space-charge region is then formed under the metal contact, which acts as a barrier to electron flow between the metal and the semiconductor. If a bias voltage V is applied to the metal (taken with respect to contact B), the metal-semiconductor barrier is found to be independent of V, hence the current $-I_0$ due to electrons flowing from the metal to the semiconductor is independent of bias. But the barrier height for electron flow from the semiconductor to the metal is lowered by an amount eV, hence the current due to electrons flowing from the semiconductor to the metal varies as $\exp(eV/kT)$. Since the currents must cancel each other for $V=0$, the characteristic is of the form

$$I_D = I_0\left[\exp\left(\frac{eV}{kT}\right) - 1\right] \tag{7.1}$$

If we define the turn-on voltage of the Schottky barrier diode as the voltage V that makes $I_D = 1$ mA, we find a turn-on voltage of about 0.40 V.

Therefore, if one connects a Schottky barrier diode between the base and the collector with the polarity as shown, and increases I_B from a zero value (Fig. 7.2b), the collector voltage V_{CE} decreases with increasing I_B, whereas the base-emitter voltage stays at about 0.70 V. When V_{CE} becomes as low as 0.30 V, the diode D has a forward bias of $0.70-0.30=0.40$ V and is turned on. The current I_D then flows directly into the collector and not into the base. If I_B is further increased, V_{CE} stays clamped at a value of 0.30 V. The currents are

$$I_B' = I_B - I_D \tag{7.2}$$

$$I_C' = \beta_F I_B' = \frac{V_{CC} - 0.30}{R_c} + I_D = \beta_F(I_B - I_D) \tag{7.2a}$$

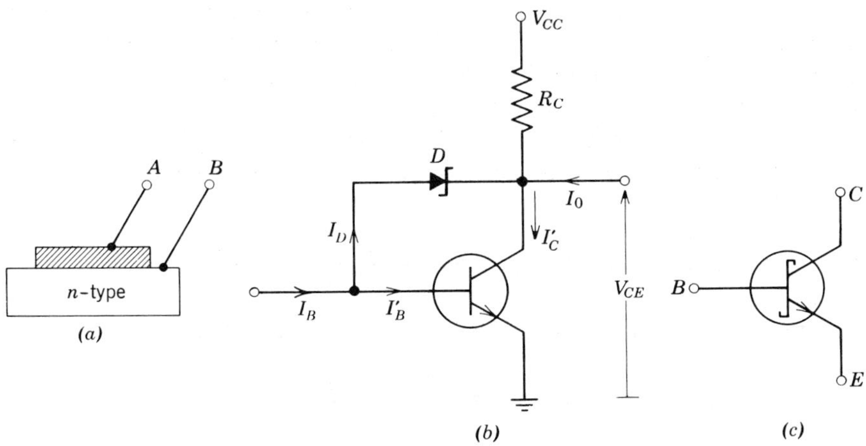

Figure 7.2. (*a*) Diagram of a Schottky barrier diode. (*b*) Transistor cricuit protected against collector saturation by a Schottky barrier diode *D* connected between the base and the collector. (*c*) Symbol of a Schottky barrier diode-protected transistor.

or

$$I_D = \frac{\beta_F I_B - (V_{CC} - 0.30)/R_c}{\beta_F + 1} \tag{7.2b}$$

$$I'_B = \frac{I_B + (V_{CC} - 0.30)/R_c}{\beta_F + 1} \tag{7.2c}$$

Therefore no excess charge is stored in the base, and no turn-off delay occurs when the base is turned off.

If an current I_0 is fed into the collector (Fig. 7.2*b*), we have instead

$$I'_B = I_B - I_D \tag{7.3}$$

$$I'_C = \beta_F I'_B = \frac{V_{CC} - 0.30}{R_c} + I_0 + I_D = \beta_F (I_B - I_D) \tag{7.3a}$$

$$I_D = \frac{\beta_F I_B - (V_{CC} - 0.30)/R_c - I_0}{\beta_F + 1} \tag{7.3b}$$

$$I'_B = \frac{I_B + (V_{CC} - 0.30)/R_c + I_0}{\beta_F + 1} \tag{7.3c}$$

The Schottky barrier diode is a majority carrier device. Therefore no lifetime effects are involved, and the turn-on and turn-off times of the device are negligibly short. The diode protector circuit thus speeds up the turn-off of the transistor considerably.

Figure 7.2*c* shows the symbol for a transistor with Schottky barrier diode protector. Note also the symbol for the Schottky barrier diode in Fig. 7.2*b*.

7.1c Transistor and MOSFET Nor and Nand Circuits

Figure 7.3*a* represents a nor circuit, since it satisfies the truth table

A	B	$Y=\overline{A+B}$
0	0	1
1	0	0
0	1	0
1	1	0

The student will recognize it as a case of wired logic.

Figure 7.3*b* represents a nand circuit, since it satisfies the truth table

A	B	$Y=\overline{AB}$
0	0	1
1	0	1
0	1	1
1	1	0

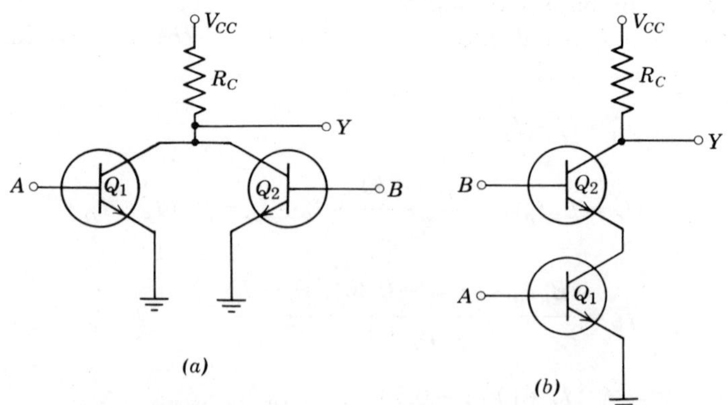

Figure 7.3. (*a*) Two-gate transistor nor circuit. (*b*) Two-gate transistor nand circuit.

These circuits have some disadvantages, however. In the first place the bases are not protected, so that large currents can flow if the base voltage is too large. For this reason the bases should be protected by series resistors, or by feeding from previous stages that provide such resistors.

The nand circuit has the additional disadvantage that, for a nand circuit with n inputs, the output voltage for a logical 0 output is $nV_{CE,\mathrm{sat}}$. This can be larger than the turn-on voltage of the next stage. Moreover, the turn-on voltage of the ith stage is $V_{BE0}+(i-1)V_{CE,\mathrm{sat}}$, and this means that the input voltages must be sufficiently large to overcome this difference in turn-on voltage.

Figure 7.4 represents an MOS *nor* and an MOS *nand* circuit. Figure 7.4*a* is analogous to Fig. 7.3*a*, hence represents a nor circuit for negative logic. Figure 7.4*b* is analogous to Fig. 7.3*b*, hence represents a nand circuit for negative logic.

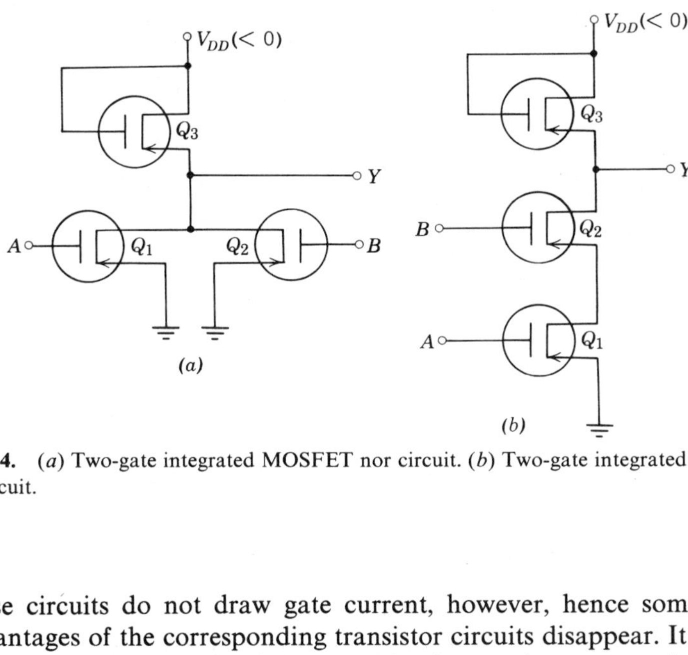

Figure 7.4. (*a*) Two-gate integrated MOSFET nor circuit. (*b*) Two-gate integrated MOSFET nand circuit.

These circuits do not draw gate current, however, hence some of the disadvantages of the corresponding transistor circuits disappear. It remains true, however, that for a nand circuit with n inputs the output voltage for a logical 0 output is nV_0, where V_0 is the output voltage of a single stage operating in the ohmic regime. Also, the turn-on voltage of the ith gate is $V_T+(i-1)V_0$. The input voltages must therefore be sufficiently large to overcome this difference in turn-on voltage.

Next we show two CMOS gates. Figure 7.5*a* represents a *positive nor** circuit. If $A=B=0$, then Q_3 and Q_4 are on and Q_1 and Q_2 are off, so $V_y=V_{DD}$. If $A=1$, then Q_4 is off and Q_1 is on, so $V_y=0$. If $B=1$, then Q_3 is off and Q_2 is on, so $V_y=0$. Therefore $V_y=0$ unless $A=B=0$, as required for a nor circuit. In a similar way it follows that Fig. 7.5*b* is a *negative nand* circuit.† For $A=B=0$, Q_1 and Q_2 are off and Q_3 and Q_4 are on, so $V_y=V_{DD}$. For $A=1$ and $B=0$, $V_y=V_{DD}$, since Q_4 is on and Q_2 is off. For $A=0$ and $B=1$, $V_y=V_{DD}$, since Q_3 is on and Q_1 is off. For $A=B=1$, Q_3 and Q_4 are off and Q_1 and Q_2 are on, so $V_y=0$. All the requirements for a negative nand are therefore satisfied.

The advantage of CMOS circuits is that they do not draw any current, hence there is almost unlimited fan-in and fan-out. Since the output voltage for a logical 0 output is zero, the turn-on voltage is $\pm V_T$, hence the noise immunity is the same.

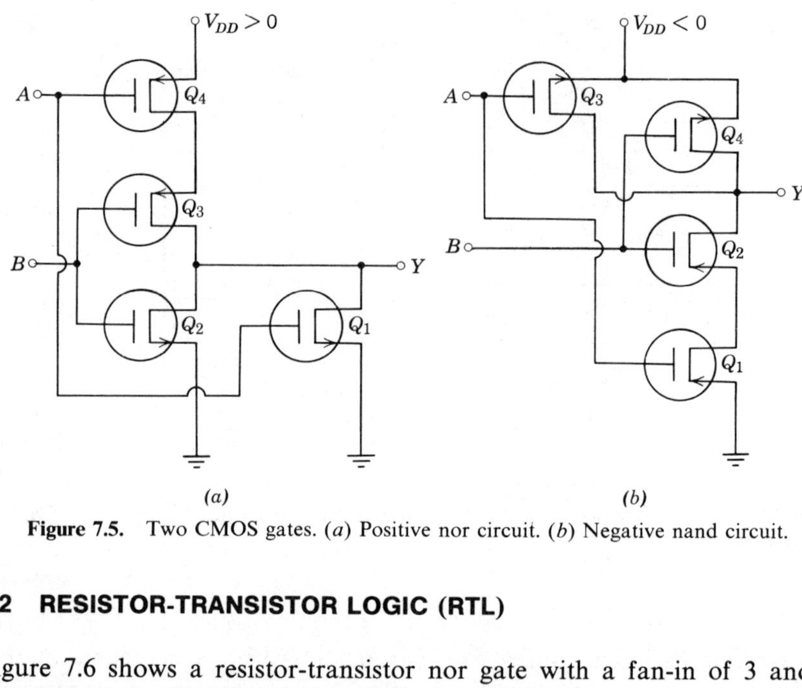

Figure 7.5. Two CMOS gates. (*a*) Positive nor circuit. (*b*) Negative nand circuit.

7.2 RESISTOR-TRANSISTOR LOGIC (RTL)

Figure 7.6 shows a resistor-transistor nor gate with a fan-in of 3 and a fan-out of 3. It should be determined of course whether such a fan-in and fan-out are allowed in a particular case.

We first investigate what voltage has to be applied to the 1 input gate when the other input gates are off. Let there be n gates and let the base

*$A=0$ means V_A close to 0 V; $A=1$ means $V_A \cong V_{DD}$ ($V_{DD}>0$, positive logic).
†$A=0$ means V_A close to 0 V; $A=1$ means $V_A \cong V_{DD}$ ($V_{DD}<0$, negative logic).

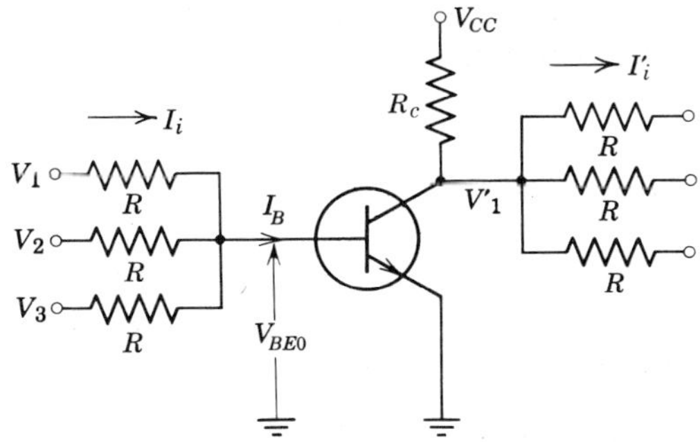

Figure 7.6 Resistor-transistor gate with a fan-in of 3 and a fan-out of 3.

current I_B be m times the saturated base current, $I_{BS}=I_{CS}/\beta_F=(V_{CC}-V_{CE,\text{sat}})/\beta_F R_c$, so the transistor is certainly saturated. We thus have

$$(n-1)\frac{V_{BE0}}{R}+mI_{BS}=\frac{V_1-V_{BE0}}{R}$$

or

$$V_1=nV_{BE0}+mI_{BS}R. \tag{7.4}$$

Example 1. $n=3$, $V_{BE0}=0.70$ V, $I_{BS}=200$ μA, $R=1000$ Ω, $m=2$, and $V_{CE,\text{sat}}=0.20$ V. Find V_1.

Answer. $V_1=3\times0.70+2\times0.20=2.50$ V.

Example 2. If in Example 1 V_1 is fed from a supply voltage V_{CC} through a resistance $R_c=1000$ Ω, find the value of V_{CC} for which $V_1=2.50$ V.

Answer. $I_1=0.70+0.70+0.40=1.80$ mA. $V_{CC}=V_1+I_1R_c=2.50+1.80=4.30$ V.

Example 3. If $R=R_c=1000$ Ω, the output feeds into three resistors R as in Fig. 7.6, and each resistor feeds into another transistor which has two other resistive gates which are off, find the value of V_{CC}.

Answer.

$$V_1'=2.50 \text{ V} \qquad \text{according to Example 1}$$

$$I_1'=1.80 \text{ mA} \qquad \text{according to Example 2}$$

$$V_{CC}-V_1'=3\times1.80\times10^{-3}\times1000;\ V_{CC}=5.40+2.50=7.90 \text{ V}.$$

These examples give some idea of the fan-in and fan-out capability of the circuit and of the values of V_{CC} needed to achieve it.

Example 4. Find the noise immunity for a 0 input and for a 1 input in Example 1 if $R_c = 1000\ \Omega$ and $V_{CC} = 5.0$ V.

Answer. For a 1 input a current $I_B = I_{BS}$ is needed to saturate the transistor. This requires $V_1 = 3 \times 0.70 + 0.20 = 2.30$ V. Now choose $V_{CC} = V_{1H}$ such that $V_1 = 2.30$ V. Since the total current is $0.70 + 0.70 + 0.20 = 1.60$ mA, we have $V_{1H} = 2.30 + 1.60 = 3.90$ V. The noise immunity for a 1 input is therefore $5.00 - 3.90 = 1.10$ V.

When $V_{BE} = V_{BE1} = 0.55$ V, base current begins to flow. In this case,

$$V_1 = nV_{BE1} = 3 \times 0.55 = 1.65 \text{ V}$$

Since $V_1 = 0.20$ V for a normal 0 input, the noise immunity for a 0 input is $1.65 - 0.20 = 1.45$ V.

One of the disadvantages of this circuit is that input gates that are off draw current. One can avoid this in two ways:

1. Making each resistor R feed into its own transistor.
2. Connecting a diode in series with each resistor. This makes the circuit a diode gate. Diode gates are discussed in Section 7.4.

We now discuss case 1 in greater detail. What we have in mind is a circuit as shown in Fig. 7.7.

It is again a nor circuit, as discussed earlier. The *fan-in capability* is unlimited; one must only require that, if only *one* transistor (e.g., Q_1) is on,

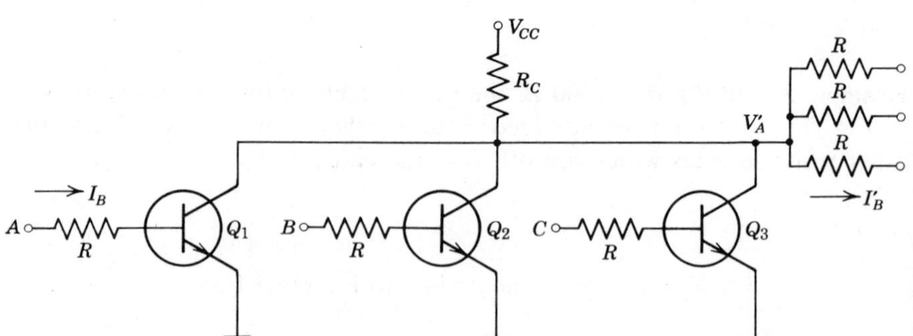

Figure 7.7. Transistor nor circuit in which each gate has its own transistor.

that it be saturated. To do so let us require $m=2$, or

$$I_B=2I_{BS}=\frac{2I_{CS}}{\beta_F}=2\frac{V_{CC}-V_{CE,\mathrm{sat}}}{\beta_F R_c} \tag{7.5}$$

and this is easily achieved. The required input voltage V_A is

$$V_A=V_{BE0}+I_BR=V_{BE0}+\frac{2(V_{CC}-V_{CE,\mathrm{sat}})R}{\beta_F R_c} \tag{7.6}$$

The spurious input pulse needed to achieve saturation for a 0 input is

$$V_A=V_{BE0}-V_{CE,\mathrm{sat}}+I_{BS}R=V_{BE0}-V_{CE,\mathrm{sat}}+\frac{(V_{CC}-V_{CE,\mathrm{sat}})R}{\beta_F R_c} \tag{7.6a}$$

The fan-out capability is limited. Suppose all transistors Q_1 through Q_3 are off and the output is a logical 1. We require that in the worst case the current I_B' flowing through each resistor R be given by Eq. (7.5). Then, for a fan-out of n and with $m=2$, the current is $2nI_{BS}$ and the voltage V_A' at the collector is given by Eq. (7.6). We then have

$$\frac{V_{CC}-V_A'}{R_c}=2nI_{BS} \qquad n=\frac{V_{CC}-V_A'}{2I_{BS}R_c} \tag{7.7}$$

Example 5. If $I_B=400\ \mu A$, $V_{BE0}=0.70$ V, $V_{CE,\mathrm{sat}}=0.20$ V, and $R=R_c=1000\ \Omega$, find $V_A'=V_A$ and the fan-out capability for $V_{CC}=5.00$ V.

Answer. $V_A'=V_{BE0}+I_BR=0.70+0.40=1.10$ V.

$$n=\frac{5.0-1.10}{2\times 0.400}=\frac{3.90}{0.80}=4.9$$

The fan-out capability is therefore 4.

Example 6. Find the height of the spurious input pulse that just produces saturation if $I_{BS}=200\ \mu A$ and $R=1000\ \Omega$, when the input is a 0.

Answer. $V_A=V_{BE0}-V_{CE,\mathrm{sat}}+I_{BS}R=0.70-0.20+0.20=0.70$ V.

7.3 DIRECT-COUPLED TRANSISTOR LOGIC (DCTL)

If the base resistors in Fig. 7.7 are omitted, one obtains DCTL. This has the advantage that the circuit can be simplified. It has the disadvantage that some of the transistors are not turned on or not turned on satisfactorily.

We illustrate the last point with the circuit in Fig. 7.8, showing two transistors driven from the same logical 1 input. The base currents are

$$I_{B1}=(1-\alpha_{F1})I_{ES1}\exp\left(\frac{eV_{BE}}{kT}\right) \qquad I_{B2}=(1-\alpha_{F2})I_{ES2}\exp\left(\frac{eV_{BE}}{kT}\right) \qquad (7.8)$$

For discrete transistors $(1-\alpha_{F1})I_{ES1}$ and $(1-\alpha_{F2})I_{ES2}$ may be sufficiently different so that I_{B1} and I_{B2} vary by one order of magnitude. The currents may then differ by so much that one of the transistors is not saturated when on. This phenomenon is called *current hogging*. For integrated transistors on the same chip much greater uniformity can be expected, and $(1-\alpha_{F1})I_{ES1}$ and $(1-\alpha_{F2})I_{ES2}$ differ by so little that the current-hogging effect can be ignored.

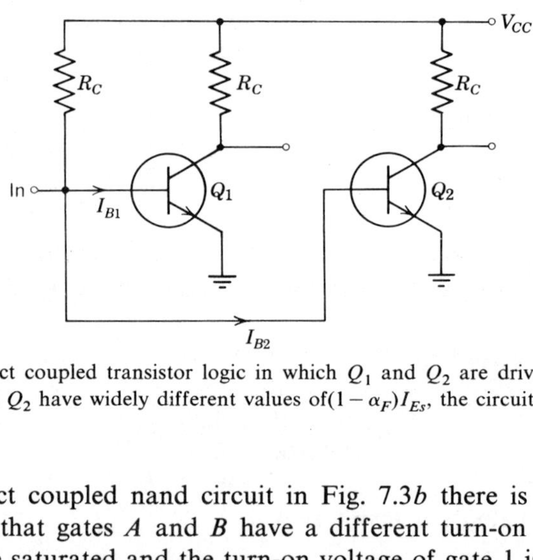

Figure 7.8. Direct coupled transistor logic in which Q_1 and Q_2 are driven from the same source. If Q_1 and Q_2 have widely different values of$(1-\alpha_F)I_{Es}$, the circuit can show current hogging.

In the direct coupled nand circuit in Fig. 7.3*b* there is the additional disadvantage that gates *A* and *B* have a different turn-on voltage. If the transistors are saturated and the turn-on voltage of gate 1 is V_{BE1} and the collector saturated voltage is $V_{CE,\text{sat},1}$, for identical transistors the turn-on voltage of gate 2 is

$$V_{BE2}=V_{BE1}+V_{CE,\text{sat},1}$$

If different gates are now driven from the same source, one must be sure that all type-A gates are driven from the same source and all type-B gates from another source. One cannot mix the gates. That is, one should not use circuits as in Fig. 7.9, because gate B has a different turn-on voltage than gate A'. It is acceptable if gates B and B' are driven from the same source and gates A and A' from another common source.

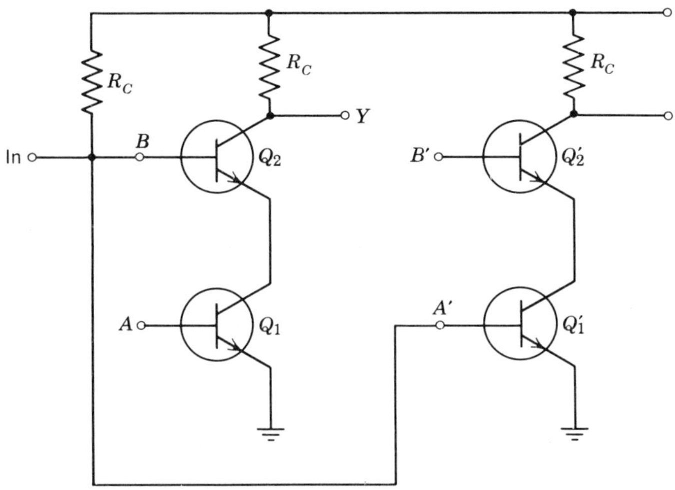

Figure 7.9. Direct coupled transistor nand circuit in which the input is connected to two unequivalent transistors.

7.4 DIODE-TRANSISTOR LOGIC (DTL)

Figure 7.10*a* shows a diode *or* circuit driving a transistor, making it a *nor* configuration. The resistor R_b is not really needed, since if Q is on, $V_{BE}=0.70$ V, independent of the value of the resistance R_b. But in fact R_b serves a useful purpose; when the transistor is being turned off, the resistance R_b draws a current V_{BE0}/R_b out of the base, and this speeds up the turn-off transient of Q.

One can even improve on this by returning the resistor R_b to a negative voltage $-V_{BB}$ ($V_{BB}>0$). The current drawn out of the base is then $(V_{BB}+V_{BE0})/R_b$, and this speeds up the turn-off transient even more. The drawback of this solution is that, when the voltage $-V_{BB}$ is present, the diode D_1 still draws current even if $V_A=0$ (Fig. 7.10*b*).

If the gates are driven from previous sources, for example, from a transistor source involving a resistance R_c that limits the flow of current,

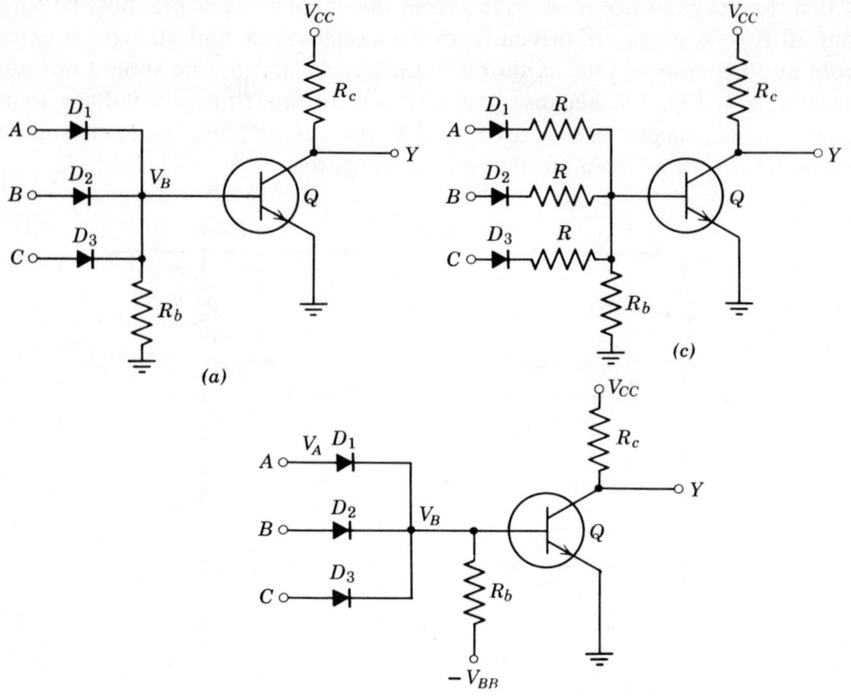

Figure 7.10 (*a*) Diode nor circuit driving a transistor. (*b*) Similar to (*a*), but the base resistance R_b is connected to a negative voltage V_{BB} to speed up the emptying of the base of Q of space charge. (*c*) Similar to (*a*), but each diode is protected by a resistor to protect the diodes and the base of Q.

the base current of Q is limited by this resistance. The turn-on voltage is then simply $2V_{BE0} \cong 1.40$ V. If this is not the case, resistors should be put in series with each diode (Fig. 7.10*c*).

The circuits in Fig. 7.10*a* and *b* are *nor* configurations. Their disadvantage is that diodes are needed to produce the nor function. This has some drawbacks if one wants to put the circuit in integrated form.

Figure 7.11*a* shows an *and* circuit driving a transistor, making it a *nand* configuration. Again the load resistor R_b is not needed for the dc operation but, since it draws a current V_{BE0}/R_b from the base when the base is being turned off, its presence speeds up the turn-off transient.

Further improvement can be obtained by returning the resistor R_b to a negative voltage $-V_{BB}$ ($V_{BB} > 0$). This is shown in Fig. 7.11*b*. It upsets the current flow in the circuit to a certain extent, however.

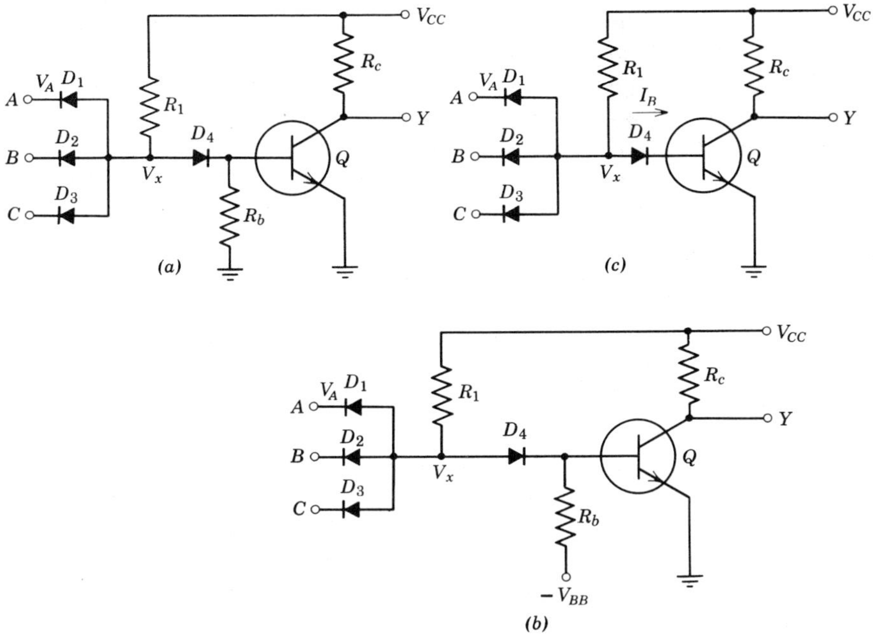

Figure 7.11 (*a*) Diode and circuit driving a transistor. The base is connected to the ground by a resistance R_b. (*b*) Similar to (*a*), but the base resistance R_b is connected to a negative voltage-V_{BB} to speedup the emptying of the base of Q of space charge. (*c*) Similar to (*a*), but resistance R_b is omitted; this circuit is used for comparing the circuit with the TTL circuit.

If $V_A=0$ V, then $V_x=0.70$ V, and a current $(V_{CC}-0.70)/R_1$ flows through R_1. If current flows through D_4, the base voltage of Q is at 0 V, and a current V_{BB}/R_b flows through R_b. The circuit must be designed so that $(V_{CC}-0.70)/R_1 > V_{BB}/R_b$, for otherwise no current would flow through D_1.

If $V_A=V_B=V_C=V_{CC}$, D_1 is off and D_4 and Q are on; the current flowing through R_1 is $(V_{CC}-1.40)/R_1$, and the current flowing through R_b is $(V_{BB}+0.70)/R_b$. Consequently the base current I_B of Q is

$$I_B=\frac{V_{CC}-1.40}{R_1}-\frac{V_{BB}+0.70}{R_b} \tag{7.9}$$

The transistor Q can be on only if $I_B>0$, or

$$\frac{V_{CC}-1.40}{R_1}>\frac{V_{BB}+0.70}{R_b} \tag{7.9a}$$

This puts constraints on V_{BB}, R_1, or R_b.

Figure 7.11*c* shows the circuit obtained when the resistor R_b is omitted. It is not useful in itself, since it has a long turn-off transient, but it is needed to explain the operation of the TTL circuit. All circuits in Fig. 7.11*a–c* are *nand* configurations.

Next something should be said about the turn-on voltage of the circuit. In Fig. 7.11*a*, when D_4 and Q are on, V_x is 1.40 V. Hence D_1 begins to be turned off if $V_A > 0.70$ V. The circuit thus has a low noise immunity. If A and B are 1's, and C is a 0, a spurious signal of about 0.85 V at gate C will turn D_3 off, hence D_4 and Q on, because the voltage across D_3 is now $1.40 - 0.85 = 0.55$ V. If V_C is driven from a previous transistor stage and that stage is on, $V_{CE,\text{sat}} \cong 0.20$ V, so the noise immunity for zero input is about 0.65 V. The noise immunity for zero input can be raised by replacing D_4 by several diodes in series.

The input diodes of the nand gates in Fig. 7.11*a* do not draw current when the inputs are 1's; when they are 0's, the diodes pass currents to the previous stage which must be designed so that it can take up this current. The maximum current passed through a diode is the current found when all the other diodes have a 1 input; the current in Fig. 7.11*a* is then

$$I_A = \frac{V_{CC} - 0.70}{R_1} \tag{7.10}$$

There is no limit to the fan-in capability, but there is a limit to the fan-out capability for the transistor Q. If it drives n different gates, each gate can, under worst-case conditions, feed a current I_A into the transistor Q. The transistor Q must then be able to take up the collector current

$$I_C = nI_A + \frac{V_{CC} - V_{CE,\text{sat}}}{R_c} \tag{7.10a}$$

without being driven out of saturation. If the transistor Q carries a base current I_B when on, the collector is saturated when $I_C < \beta_F I_B$. The condition for n is therefore

$$\beta_F I_B > nI_A + \frac{V_{CC} - V_{CE,\text{sat}}}{R_c} \tag{7.10b}$$

This limits the fan-out.*

To improve the fan-out capability, the input diodes D_1 through D_3 can be replaced by *p–n–p* transistors in a common emitter connection, as

*Often an even more serious limitation is due to the collector series resistance r_c (see Section 3.2b).

shown in Fig. 7.12. Again, a resistance R connects the base of Q_4 to the ground, to decrease the turn-off time.

Since each common emitter circuit has a current amplification factor (β_F+1), the maximum current that flows out of input A when a 0 is applied is

$$I_A = \frac{V_{CC}-0.70}{(\beta_F+1)R_1} \tag{7.11}$$

The fan-out capability can therefore be increased by a factor of β_F+1.

This is easily achieved in practice if the integrated transistors are triple-diffused transistors. Here one starts with an n-type epitaxial layer on a p-type substrate. One first diffuses p-regions all the way through, to make isolated n-type islands separated by p-type walls. Into these islands one diffuses a p-type base and an n-type emitter, and what is left of the island is used as a collector. There are three diffusion steps involved, hence the name "triple-diffused" transistor.

One now uses the base as the emitter, the collector as the base, and the substrate as the collector. One can even omit the last n-type diffusion altogether, since one wants a p–n–p transistor anyway. But since the

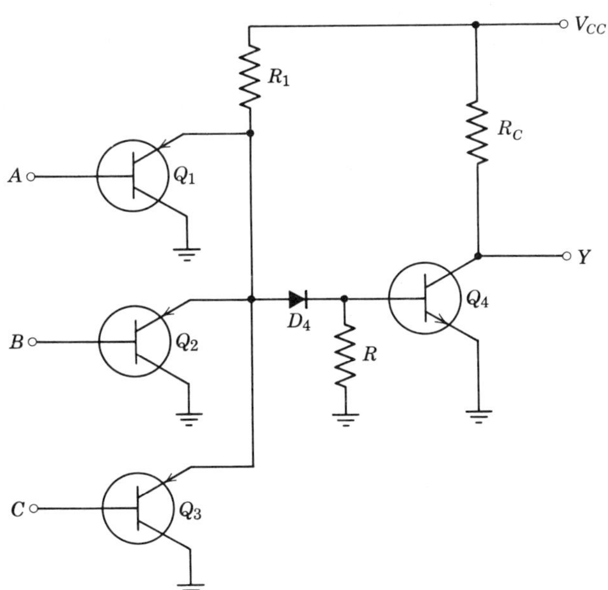

Figure 7.12. Three-input nand gate in which each input diode is replaced by a p-n-p transistor in a common emitter connection.

various transistors have the substrate in common, they can be used only in the emitter follower connection. We saw, however, that this is no handicap.

The *p–n–p* transistors thus introduced are not especially built for this purpose; therefore β_F is relatively small, often only about 3. Since the emitter follower has a current gain $1+\beta_F$, this corresponds to a gain of about 4.

The nature of the circuit is the same as that of the diode circuit in Fig. 7.11*a*; therefore Fig. 7.12 represents a *nand* configuration.

7.5 THE TRANSISTOR-TRANSISTOR LOGIC GATE (TTL)

Figure 7.13 shows a TTL gate driving a transistor Q_1. It consists of a transistor with two or more emitters (here we use three), designated by the symbol T_1, with the collector feeding into the base of the transistor Q_1.

Before discussing the operation of the circuit, we first point out the similarity of the configurations in Figs. 7.13 and 7.11*c*. The three emitter diodes compare to diodes D_1 through D_3, and the collector diode corresponds to diode D_4. There is an important difference, however, since the device T_1 can operate as a transistor under certain conditions; as we shall see in a moment, this decreases the turn-off time of the circuit considerably.

We now discuss the operation of the circuit. If $V_A = V_B = V_C \cong V_{CC}$, the emitters are off and T_1 acts as an inverted transistor with

$$I_B = \frac{V_{CC} - 1.40}{R_b} \qquad I_E = -\beta_R I_B \qquad I_C = (1+\beta_R)I_B = I_{B1}$$

We assume in the following that T_1 has $\beta_R \cong 0$.

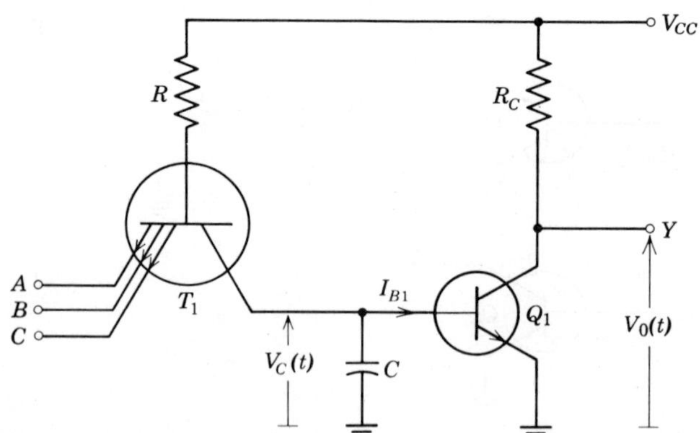

Figure 7.13. TTL gate. The capacitance C represents the capacitance between the collector and the substrate, plus the stored charge capacitance of the transistor Q_1.

The current is sufficiently large to drive the transistor Q_1 to saturation, so $V_0(t) \cong 0$. Therefore a 1 input at each gate gives 0 output for Q_1.

If one of the gate voltages is a 0, the gate T_1 acts as a normal transistor, discharging the capacitance C until $V_{CE} \cong 0$; C represents the input capacitance of Q_1 plus a possible capacitance between the collector and the substrate. That is, Q_1 will now be turned off, so a 0 input at either one, two, or all the emitters gives a 1 output. Only if all the input terminals are fed by 1's does one obtain a 0 output. The circuit is therefore a *nand* circuit.

The turn-off transient is short. If $V_A = V_B = V_{CC}$ for $t<0$, and $V_C(t)$ switches to zero at $t=0$, the voltage $V_c(t)$ on the capacitor C has the value $V_c(0^+) = V_{BE0}$ at $t=0^+$. The current with which the capacitor C is discharged is $\beta_F I_B$. Since the total charge on the capacitor is CV_{BE0}, the discharge time is

$$t_d = \frac{CV_{BE0}}{\beta_F I_B} \tag{7.13}$$

The turn-off time t_2 of Q_1 is even considerably shorter than t_d, since Q_1 is already turned off after the base-emitter voltage of Q_1 drops by as little as 0.15 V.

We now investigate the turn-on time t_1 of Q_1. Let $V_B = V_C = V_{CC}$ and $V_A = 0$ for $t<0$ and let V_A be switched to V_{CC} at $t=0$. Then $V_c(0^+) \cong 0$. The transistor Q_1 is turned on as soon as $V_c(t)$ reaches the value V_{BE0}. The current I_B thus gradually drops from the value $(V_{CC} - V_{BE0})/R_b$ at $t=0$ to the value $(V_{CC} - 2V_{BE0})/R_b$ at $t=t_1$. Neglecting this small difference, and bearing in mind that the total charge on C at $t=t_1$ is CV_{BE0}, we have

$$t_1 \cong \frac{CV_{BE0}}{I_B} \tag{7.14}$$

which is much longer than t_d.

Example 3. Find t_1 and t_d if $C=50$ pF, $I_B=1$ mA, $\beta_F=50$, and $V_{BE0}=$ 0.70 V.

Answer.

$$t_1 = \frac{CV_{BE0}}{I_B} = \frac{50\times 10^{-12}\times 0.70}{10^{-3}} = 3.5\times 10^{-8}\text{ s}$$

$$t_d = \frac{CV_{BE0}}{\beta_F I_B} = \frac{3.5\times 10^{-8}}{50} = 0.7\times 10^{-9}\text{ s}$$

7.6 TOTEM-POLE CIRCUITS

Totem-pole circuits combine inverter and emitter follower circuits. The resulting combination has the properties of an inverter circuit with small turn-on and turn-off times and with a high current capability.

7.6a Direct-Drive Totem-Pole Circuit

Figure 7.14 shows a direct-drive totem-pole circuit. A logical 0 at the input turns Q_1 and Q_2 off. This means that Q_3 has a base current I_{B3} flowing through the resistor R_2, which turns Q_3 on and makes it act as an emitter follower. Consequently the output voltage V_0 is high, and the circuit gives a logical 1 output. A logical 1 at the input drives Q_1 and Q_2 to saturation and turns Q_3 off. Therefore V_0 is close to 0 V, and the output is a logical 0.

To understand the turn-off time of the output we switch V_i on at $t=0$. Q_1 and Q_2 are then on and saturated, and the output voltage V_0 drops to zero. If both transistors are identical,

$$I_{B2}=\frac{V_i-V_{BE0}}{2R_1} \tag{7.15}$$

hence, as long as the transistor is nonsaturated,

$$I_{C2}=\beta_F I_{B2} \tag{7.16}$$

which is quite large. The current I_{C2} discharges the load capacitance C_0, representing subsequent stages, rapidly to practically 0 V. Therefore the

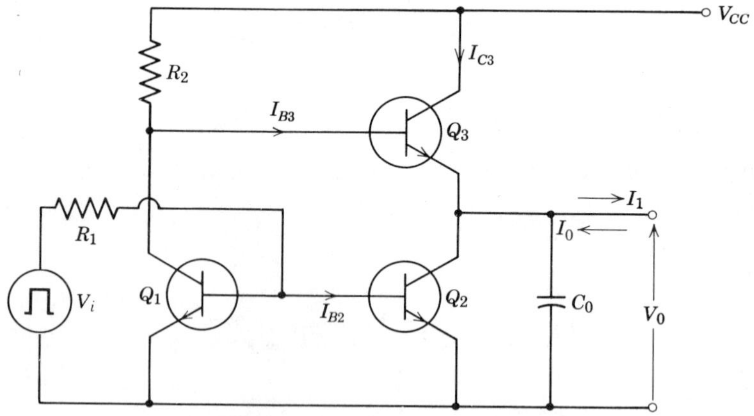

Figure 7.14. Direct-drive totem=pole circuit.

turn-off time is

$$t_2 \cong \frac{C_0 V_0}{I_{C2}} \tag{7.17}$$

To understand the turn-on time t_1 of the output, we switch V_i off at $t=0$. Q_1 and Q_2 are then off, and

$$I_{B3} = \frac{V_{CC} - V_0 - V_{BE0}}{R_2} \tag{7.18}$$

can be quite large. Since Q_3 acts as an emitter follower, the time constant τ of the output circuit is

$$\tau = \frac{C_0 R_2}{1+\beta_{F3}} \tag{7.19}$$

since C_0 "sees" a resistance $R_2/(1+\beta_{F3})$. The actual turn-on time is defined as the time needed for 90% completion of the transient. This yields

$$t_1 = 2.3\tau \tag{7.19a}$$

Example 1. If $C_0 = 50$ pF, $V_0 = 5$ V, $I_{C2} = 100$ mA, $R_2 = 2000$ Ω, and $\beta_{F3} = 100$, find t_2 and t_1.

Answer.

$$t_2 = \frac{C_0 V_0}{I_{C2}} = \frac{50\times 10^{-12}\times 5}{0.10} = 2.5\times 10^{-9} \text{ s}$$

$$t_1 = 2.3\times \frac{50\times 10^{-12}\times 2\times 10^{3}}{100} = 2.3\times 10^{-9}$$

Next something should be said about the output current capability of the circuit. Suppose, when the output is at a logical 0, that the gates into which the output feeds force a current I_0 into the collector of Q_2. Then Q_2 remains saturated as long as

$$I_0 \leqslant \beta_{F2} I_{B2} \tag{7.20}$$

By proper circuit design, this can be a large current.

Suppose, when the output is at a logical 1, that the gates into which the output feeds draw a current I_1 from the output. Obviously, the maximum current that can be drawn is

$$I_1 = (\beta_{F3} + 1) I_{B3} \tag{7.21}$$

but this does not solve our problem, since I_{B3} depends on V_0. We observe that the voltage V_0 adjusts itself so that

$$I_{B3} = \frac{V_{CC} - V_{BE0} - V_0}{R_2} = \frac{I_1}{\beta_{F3} + 1} \tag{7.22}$$

or

$$V_0 = V_{CC} - V_{BE0} - \frac{I_1 R_2}{\beta_{F3} + 1} \tag{7.22a}$$

Even for relatively large values of I_1 this can be a logical 1 output, since $R_2/(\beta_{F3}+1)$ can be a small resistance.

The current capability given by Eq. (7.20) often cannot be met in practice, because of the effect of the collector series resistance r_{c2}. Because of this series resistance, the output voltage of Q_2 can be written as

$$V_{CE2} = V_{CE,\text{sat}} + I_0 r_{c2} \tag{7.22b}$$

If this voltage is too large, the output no longer presents a 0. For example, if the circuit feeds into similar circuits, the circuits are on when $V_{CE2} > V_{BE0}$. In practice one may want to tolerate $V_{CE2} = 0.50$ V. For a given r_{c2} this limits I_0 to a value far below the one given by Eq. (7.20).

Example 1. If $r_{c2} = 10\ \Omega$ and $V_{CE,\text{sat}} = 0.20$ V, find the value of I_0 that gives $V_{CE2} = 0.50$ V.

Answer. $0.50 = 0.20 + 10 I_0$, or $I_0 = 30$ mA.

The collector series resistance r_{c3} of Q_3 has no effect as long as Q_3 is not saturated; the value of V_{CE3} is then immaterial. Note, however, that Q_3 saturates when $V_{CC} - I_{C3} r_{c3} = V_0 + V_{CE,\text{sat}}$. This puts a limitation on I_{C3}, and hence on I_1, for any given V_0.

7.6b Phase Splitter-Drive Totem-Pole Circuit

This circuit, which is shown in Fig. 7.15, is similar to the previous circuit. Q_1 acts as an emitter follower driving Q_2. If V_i is at a logical 0, Q_1 and Q_2

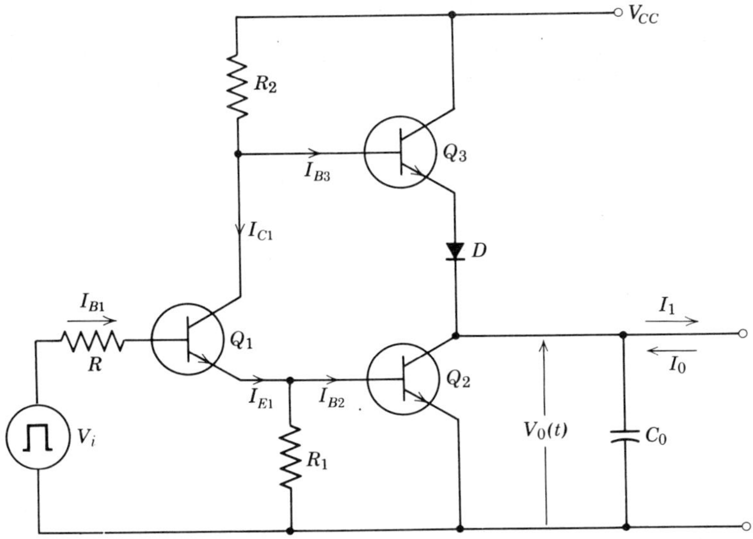

Figure 7.15. Phase splitter-drive totem = pole circuit.

are off, whereas Q_3 is on and acts as an emitter follower; therefore V_0 is at a logical 1. The only difference compared to the previous circuit is that there is a voltage drop of 0.70 V in diode D.

If V_i is a logical 1, Q_1 is on and saturated, driving Q_2 to saturation. To achieve this, V_i must be larger than two diode voltage drops, or 1.40 V. If I_{B1} is the base current of Q_1, the emitter current I_{E1}, practically all flowing into the base of Q_2 is,

$$I_{E1} = \frac{V_{CC} - V_{BE0} - V_{CE,\text{sat}}}{R_2} + I_{B1} \tag{7.23}$$

hence the output current capability is given by

$$I_0 \leqslant \beta_{F2} I_{B2} = \beta_{F2}\left(\frac{V_{CC} - V_{BE0} - V_{CE,\text{sat}}}{R_2} + I_{B1} - \frac{V_{BE0}}{R_1}\right) \tag{7.24}$$

This is larger than in the previous case. Whether or not this is useful depends on the value of the collector series resistance r_{c2} of Q_2.

If V_i is at a logical 0, Q_1 is off and the voltage drop across R_1 is zero, hence Q_2 is off. Q_3 now acts as an emitter follower, having a current capability similar to that in the previous case, except that the voltage drop in diode D must be taken into account. Diode D is needed, because otherwise Q_3 can draw current when Q_1 and Q_2 are saturated ($V_{BE3} \cong$ 0.70 V when diode D is not present).

The turn-on and turn-off times of the output circuit are similar to the ones in the previous case. The input circuit of Q_1 must be protected so that I_{B1} does not become too large. This can be done by driving it from a TTL gate or from a diode and gate.

7.6c TTL Gates with Schottky Barrier Protection Diodes

Figure 7.16 shows a TTL gate driving a transistor; both devices are protected by a Schottky barrier diode. To understand its operation we consider two cases:

1. *A*, *B*, and *C all high*. $V_A \cong V_B \cong V_C \cong V_{CC}$.

Current flows through R_b and the Schottky barrier diode (practically no current flows through the collector diode, since it needs 0.70 V to be turned on and the Schottky barrier diode needs only 0.40 V). Hence

$$I_B = \frac{V_{CC} - 0.40 - 0.70}{R_b}$$

Q is on and not saturated, and $V_y = 0.70 - 0.40 = 0.30$ V.

2. *A low; B, and C high*. $V_A \cong 0$; $V_B \cong V_C \cong V_{CC}$.

The current I_B flows through R_b and the Schottky barrier diode and returns to A via the collector (a small part of I_B flows into the base), so T acts more or less as a transistor. The bias between the collector and the

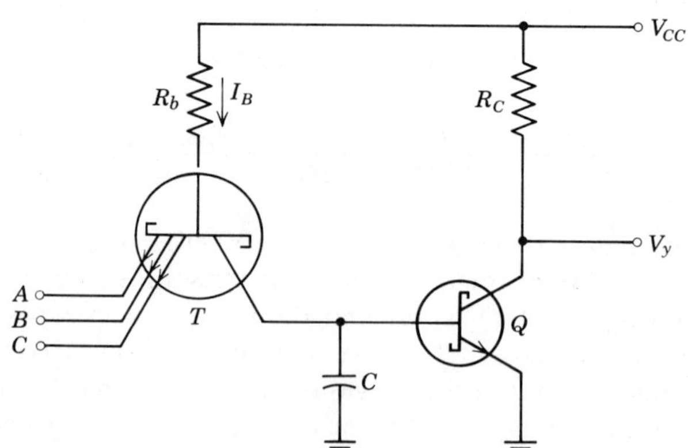

Figure 7.16. Schottky barrier-protected TTL circuit.

emitter A is therefore 0.30 V, and

$$I_B = \frac{V_{CC} - 0.40 - 0.30}{R_b}$$

Q is now off of course, because the base-emitter voltage of Q is only 0.30 V if $V_A = 0$ V.

The advantages of this circuit are:

1. The transistor Q is not saturated, so there is no turn-off delay.

2. The voltage V_{BE} of Q only has to drop from 0.70 V to 0.30 V when Q is turned off, and to rise from 0.30 V to 0.70 V when Q is turned on; therefore the turn-on and turn-off times are shorter.

7.6d Totem-Pole Circuit with Schottky Barrier-Protected Transistors

Figure 7.17 shows a totem-pole circuit with a reasonable current capability for a 1 output. It is achieved by using two emitter followers in series. We now explain the circuit in some detail.

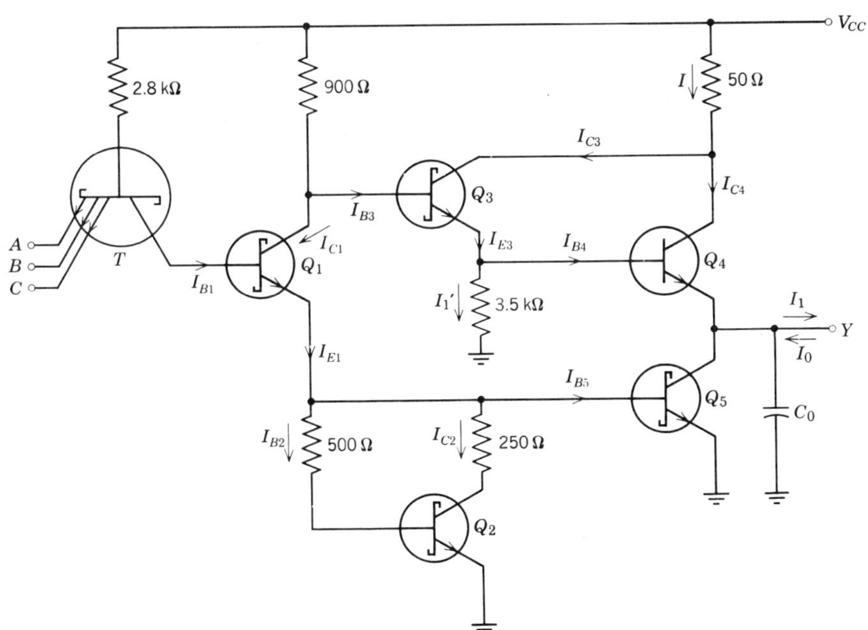

Figure 7.17. More complicated Schottky barrier-protected circuit feeding into a totem-pole driver circuit. Note that Q_4 is not protected, since it cannot be saturated. The function of Q_2 is mainly to empty the base of Q_5 of stored charge when Q_5 is turned off.

Let all transistors have $\beta_F = 100$, $V_{BE0} = 0.70$ V, and $V_{CE} = 0.30$ V when the Schottky barrier diodes are operating. We now consider two cases:

1. *V_A, V_B, and V_C all high.* $V_A \cong V_B \cong V_C \cong V_{CC} = 5.0$ V.

Q_1, Q_2, and Q_5 are on, and so is the Schottky barrier diode of T. Hence

$$I_{B1} = \frac{5.00 - 2\times 0.70 - 0.40}{2800} = \frac{3.20}{2800} = 1.14 \text{ mA}$$

$$I_{C1} = \frac{5.00 - 0.70 - 0.30}{900} = 4.44 \text{ mA} \qquad I_{E1} = I_{B1} + I_{C1} = 5.6 \text{ mA}$$

We assume I_{B2} to be relatively small since Q_5 has $V_{BE} \cong 0.70$ V. The main function of Q_2 is to empty the base of Q_5 when the circuit is turned off. $I_{C2} = (0.70 - 0.30)/250 = 1.60$ mA. Consequently $I_{B5} \cong I_{E1} - I_{C2} = 4.0$ mA, and the current capability of Q_5 is approximately 400 mA for a 0 output.

2. *V_A low; V_B, and V_C high.* $V_A \cong 0$; $V_B \cong V_C \cong V_{CC} = 5.0$ V.

Q_1, Q_2, and Q_5 are now off, and the base current of T flows through the emitter A. Q_3 and Q_4 are on. Because of the voltage drop in the 50-Ω resistor V_{CE3} is clamped to 0.30 V. But Q_4 is not saturated.

We now try to find the current capability of Q_4 under the assumption that $V_y = 3.0$ V. The base voltage of Q_4 is therefore 3.70 V, and the base voltage of Q_3 is 4.40 V. Hence $I_{B3} = (5.00 - 4.40)/900 = 0.67$ mA. The collector voltages of Q_3 and Q_4 are thus $3.7 + 0.3 = 4.0$ V above ground, or $V_{CE4} = 1.0$ V, so Q_4 is not saturated. Hence the current through the 50-Ω resistor is

$$I = I_{C3} + I_{C4} = \frac{5.00 - 4.00}{50} = 20.0 \text{ mA}$$

Now $I_{E3} = I_{C3} + I_{B3} = I_{C3} + 0.67$ mA. The current I_1' flowing through the 3.5-KΩ resistor is

$$I_1' = \frac{3.70}{3.50} = 1.06 \text{ mA}$$

Hence $I_{B4} = I_{E3} - I_1' = I_{C3} + 0.67 - 1.06 = I_{C3} - 0.39$. But $I_{C4} = 100\, I_{B4}$, so

$$I = I_{C3} + 100 I_{B4} = 20 \text{ mA}$$

or

$$I_{C3} + 100 I_{C3} - 39 = 20.0 \text{ mA}$$

$$I_{C3} = \frac{59}{101} = 0.584 \text{ mA} \qquad I_{C4} = 20.0 - 0.584 = 19.4 \text{ mA}$$

$$I_{B4} = \frac{I_{C4}}{100} = 0.194 \text{ mA} \qquad I_1 = 19.4 + 0.194 = 19.6 \text{ mA}$$

The current capability of Q_4 rises considerably if V_y is lowered. It also rises when the 50-Ω resistor is made smaller, since it limits I_1.

We finally calculate the turn-on time of Q_4. We observe that $I_{E3} = I_{C3} + 0.67 = 0.584 + 0.67 = 1.25$ mA, so the resistance "seen" when looking into the emitter is

$$r_{e3} = \frac{kT}{eI_{E3}} = \frac{25.8\times10^{-3}}{1.25\times10^{-3}} = 21\ \Omega$$

Since $I_{B4} = 0.194$ mA,

$$r_{\pi4} = \frac{25.8\times10^{-3}}{0.194\times10^{-3}} = 133\ \Omega$$

When the series base resistance is neglected, the total resistance is therefore 154 Ω, and the resistance "seen" by the emitter side of Q_4 is $154/(1+\beta_F) \cong 1.53$ Ω. If $C_0 = 100$ pF, the time constant is 1.53×10^{-10} s and the turn-on time is 2.3 times as large, or 3.5×10^{-10} s for a 0% to 90% response. The turn-on time is thus extremely short.

7.7 EMITTER-COUPLED LOGIC (ECL)

In emitter-coupled logic the transistors are never saturated. Hence phenomena like turn-off delay cannot occur. Moreover, one essentially switches *emitter current*, and this goes faster than switching base current, since a much lower impedance level is involved. To understand the operation of the circuit, we turn to Fig. 7.18. It shows a differential amplifier, driven from a constant-current source, and a reference voltage V_R. If $V_S = V_R$ and $R_s = 0$, half the current I_0 flows through each transistor. If V_S is about 0.12 V higher than V_R, all the current flows through Q_1 and, if V_S is about 0.12 V lower than V_R, all the current flows through Q_2 (Section 5.5). In the first case $V_y = V_{CC}$ (1 output), and in the second case $V_y = V_{CC} - I_0R_c$ (0 output). By choosing R_c one can make V_y such that the transistor never saturates. A relatively small voltage difference can thus perform the switching. This switching operation is called *current-mode switching.*

In the switching process the total current does not change, hence

$$I_{C1} + I_{C2} = I_0 \tag{7.25}$$

But, according to the charge-control equations, if $\tau_\pi = \tau_{BF}$

$$I_{B1} = \frac{dq_F}{dt} + \frac{q_F}{\tau_{BF}} \qquad I_{C1} = \beta_F\frac{q_F}{\tau_{BF}} \tag{7.26}$$

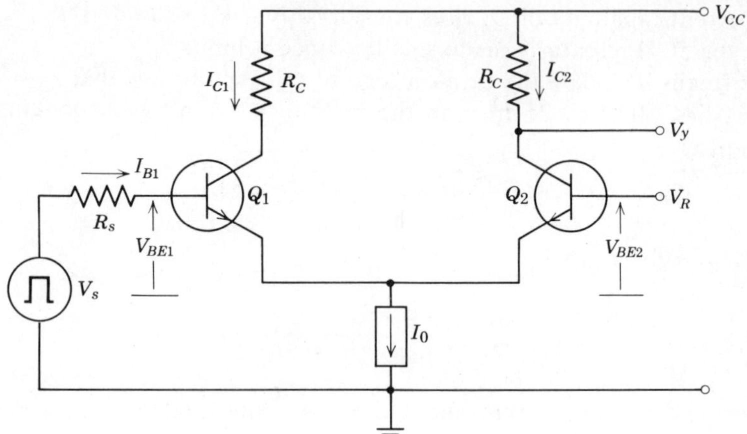

Figure 7.18. Differential transistor switching circuit to explain the fast switching of ECL.

Since the switching goes very fast, dq_F/dt is much larger than q_F/τ_{BF}, hence in good approximation

$$I_{B1}=\frac{dq_F}{dt}=\frac{\tau_{BF}}{\beta_F}\frac{dI_{C1}}{dt}=\tau_F\frac{dI_{C1}}{dt} \qquad \text{where } \tau_F=\frac{\tau_{BF}}{\beta_F} \tag{7.27}$$

and, similarly,

$$I_{B2}=\tau_F\frac{dI_{C2}}{dt} \tag{7.28}$$

Since $I_{C1}+I_{C2}=I_0$, it follows that, during switching,

$$I_{B1}=-I_{B2} \tag{7.29}$$

If V_{BE1} and V_{BE2} are the emitter-base voltages,

$$V_S-V_R=V_{BE1}-V_{BE2}+I_{B1}R_S \tag{7.30}$$

Finally,

$$I_{C1}=\alpha_F I_{ES}\left[\exp\left(\frac{eV_{BE1}}{kT}\right)-1\right] \tag{7.31}$$

$$I_{C2}=\alpha_F I_{ES}\left[\exp\left(\frac{eV_{BE2}}{kT}\right)-1\right] \tag{7.32}$$

This set of equations is difficult to solve; when they are solved numerically, it is found that the 10–90% rise time of I_{C1} is about an order of magnitude *larger* than τ_F. If $\beta_F \cong 100$, it is thus seen that the rise time is about one order of magnitude *smaller* than $\tau_{BF} = \tau_\pi$. Current-mode switching is therefore much faster than switching the base current of a transistor on and off; moreover, it has no Miller effect.

Another interesting feature is that ECL can give rise to either or or nor circuits, depending on the collector from which one takes the output. Moreover, if wired logic is used, the logic is not always of the and type, that is, if Y_1 and Y_2 are the outputs, the output Y for the wired logic is not necessarily $Y_1 Y_2$.

We now discuss these aspects in some detail. Figure 7.19 shows an ECL circuit. Here R_1 and R_2 are quite small (a few hundred ohms), and R_e and R_0 are much larger. There are two inputs A and B, a reference electrode of voltage V_{ref}, and two emitter followers with outputs Y and Y'. The combination A–B–Y gives an *or* circuit, and the combination A–B–Y' gives a *nor* circuit, both for positive logic. Note that here the 0 level does not lie at 0 V. Usually $V_{\text{ref}} \cong [(V_4)_{\text{max}} + (V_4)_{\text{min}}]/2$ and $(V_A)_{\text{max}} = (V_4)_{\text{max}}$.

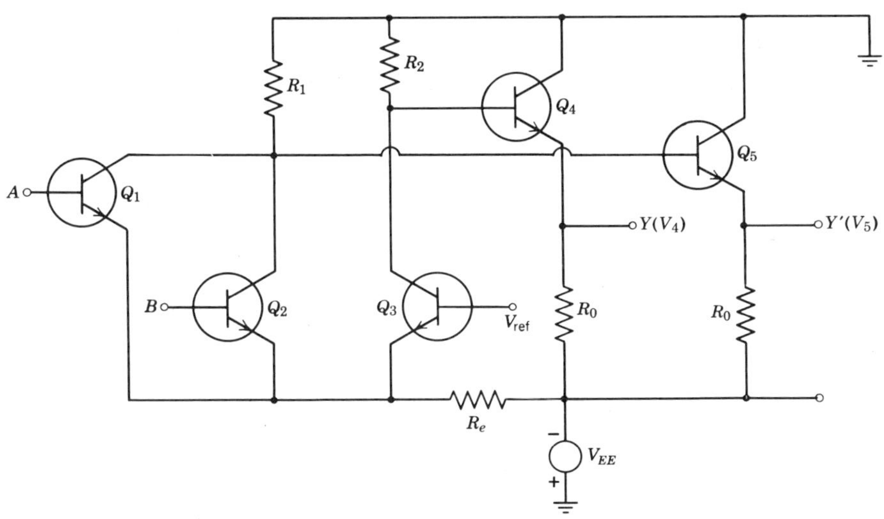

Figure 7.19. ECL circuit with two outputs $Y(V_4)$ and $Y'(V_5)$, giving an or and a nor circuit, respectively.

To show this we observe that, if the voltages V_A and V_B are both low, all current flows through Q_3, hence V_4 is low (0 output). If A, B, or both are high, no current flows through Q_3, hence V_4 is high (1 output). Hence the A–B–Y combination gives an *or* operation. For the same reason V_5 is high

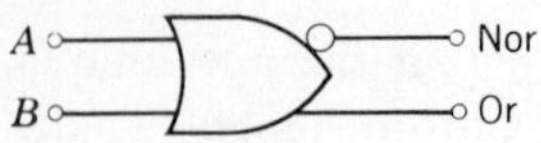

Figure 7.20. Symbol for an ECL circuit.

if V_A and V_B are both low, and V_5 is low otherwise. Hence the $A–B–Y'$ combination gives a *nor* operation. The symbol for the circuit is shown in Fig. 7.20. If $I_{C1}R_1 = I_{C3}R_2$, both outputs have the same voltage swing.

To demonstrate the statement about wired logic we redraw the circuit in Fig. 7.21. We now have the truth table

A	B	$V_1(A+B)$	$V_2(\overline{A+B})$
0	0	$V_{CC} - I_E R_C$	V_{CC}
1	0	V_{CC}	$V_{CC} - I_E R_C$
0	1	V_{CC}	$V_{CC} - I_E R_C$
1	1	V_{CC}	$V_{CC} - I_E R_C$

If we tie the outputs Y_a and Y_b of two gates together and denote the new output by Y, we have the truth table

Y_a	Y_b	V_y
0	0	$V_{CC} - I_E R_C$ (current flows through both outputs)
0	1	$V_{CC} - \frac{1}{2} I_E R_C$ (current flows through one output)
1	0	$V_{CC} - \frac{1}{2} I_E R_C$ (current flows through one output)
1	1	V_{CC} (no current in either output)

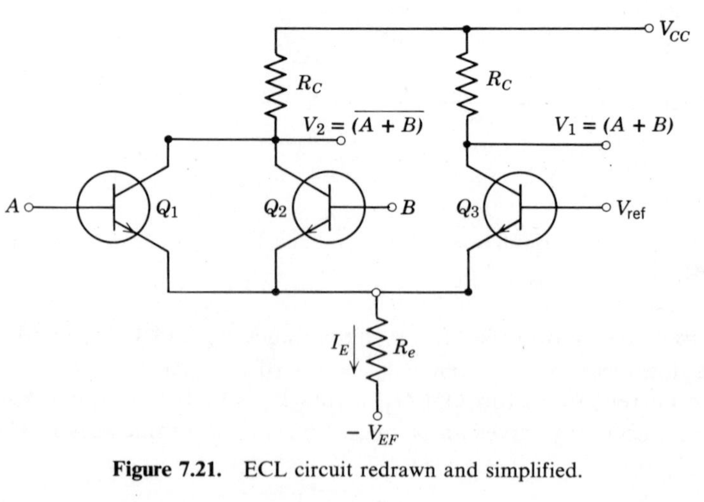

Figure 7.21. ECL circuit redrawn and simplified.

If we now take the borderline between 0 and 1 at $V_{CC} - \frac{3}{4} I_E R_c$, the output is $Y_a + Y_b$, since

Y_a	Y_b	Y
0	0	0
1	0	1
0	1	1
1	1	1

which is a *wired or*. If, however, we take our borderline between 0 and 1 at $V_{CC} - \frac{1}{4} I_E R_c$, the output is $Y_a Y_b$

Y_a	Y_b	Y
0	0	0
1	0	0
0	1	0
1	1	1

and this is a *wired and*.

7.8 MOS DRIVER CIRCUITS

Here we discuss two MOS driver circuits used to charge or discharge a load capacitance C_d rapidly.

Figure 7.22 shows a push-pull driver circuit. We use *n*-channel devices in this example. To understand the operation of the circuit, we consider first the steady-state outputs for low- and high-input voltages and then investigate transient response.

1. *Input low* (0 *input*).

Q_1 and Q_4 are off; V_A and V_0 rise until Q_2 and Q_3 do not draw current. Since Q_2 is always saturated, V_A rises until

$$I_{D2} = K_2 (V_{GG} - V_{T2} - V_A)^2 = 0 \tag{7.33}$$

so $V_A = V_{GG} - V_{T2}$ in the final state.

Suppose Q_3 is also saturated. Then, ultimately,

$$I_{D3} = K_3 (V_A - V_{T3} - V_0)^2 = 0 \tag{7.34}$$

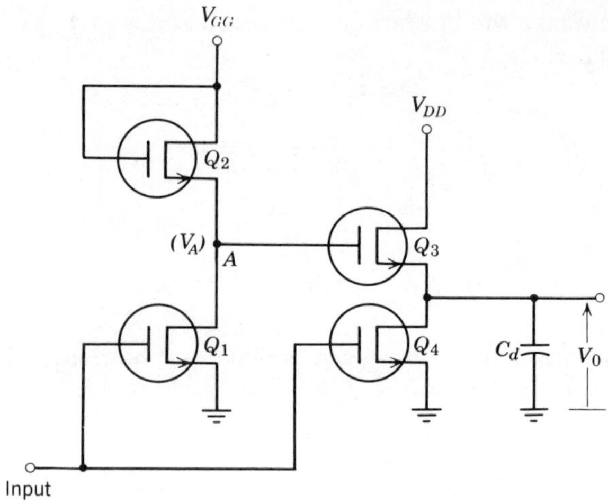

Figure 7.22. Push-pull MOSFET driver circuit. (From *MOS Integrated Circuitts*, edited by W. Penney and L. Lau, © 1972 by Litton Educational Publishing, Inc. Reprinted by permission of Van Nostrand Reinhold Company.).

so, at a steady state,

$$V_0 = V_A - V_{T3} = V_{GG} - V_{T2} - V_{T3} \tag{7.34a}$$

This situation occurs when

$$V_{DD} \geqslant V_{GG} - V_{T2} - V_{T3} \tag{7.34b}$$

If this is not the case, V_0 rises to the steady-state value V_{DD}.

2. *Input high* (1 *input*).

Q_1 and Q_4 are on, V_A is low, of the order of $V_{on} < V_{T3}$, and $I_{D1} = I_{D2}$ in the final state. Therefore Q_3 is off, and Q_4 discharges C_d until V_0 reaches 0 V.

The output voltage thus swings from $V_{GG} - V_{T2} - V_{T3}$ or V_{DD}, whichever is smaller, to 0 V.

We now turn to the transient response.

1. *Input switches from low to high.*

Q_1 and Q_4 are turned on immediately, and V_A and V_0 drop until V_A reaches the value V_{on} and V_0 reaches 0 V. This operation is materially helped if V_A drops much more rapidly than V_0; since Q_3 was off just before switching, it stays off. As a consequence C_d is discharged rapidly. This desirable feature is obtained by proper dimensioning of the circuit.

2. *Input switches from high to low.*

Q_1 and Q_4 are turned off immediately, and V_A rises. This causes Q_3 to be turned on and makes V_0 rise. This operation is materially helped if V_A rises rapidly, because this turns on Q_3 more strongly and makes C_d charge rapidly to the value $V_{GG} - V_{T2} - V_{T3}$ or V_{DD}, whichever is smaller. This desirable feature is obtained by proper dimensioning of the circuit.

Figure 7.23 shows a bootstrapped push-pull circuit made by connecting the gate of Q_2 to V_{GG} via a MOSFET load Q_5 and adding a small capacitance C_2 of about 1 pF between the gate and the source of Q_2. Q_5 can be a minimum-size device, since its only function is to charge the capacitor C_2. The operation of this circuit can be understood as follows.

1. *Input switches from high to low.*

Just before switching $V_B = V_{GG} - V_{T5}$, $V_A = V_{on}$, and $V_{on} = 0$ V. At switching Q_1 and Q_4 are turned off immediately and, since Q_2 is on, V_A rises rapidly. This keeps Q_5 off, hence prevents discharging of C_2. As a consequence V_B rises with V_A (bootstrap operation), so $V_{GS2} = V_B - V_A$ remains practically constant. The rapid rise of V_A continues until V_A reaches

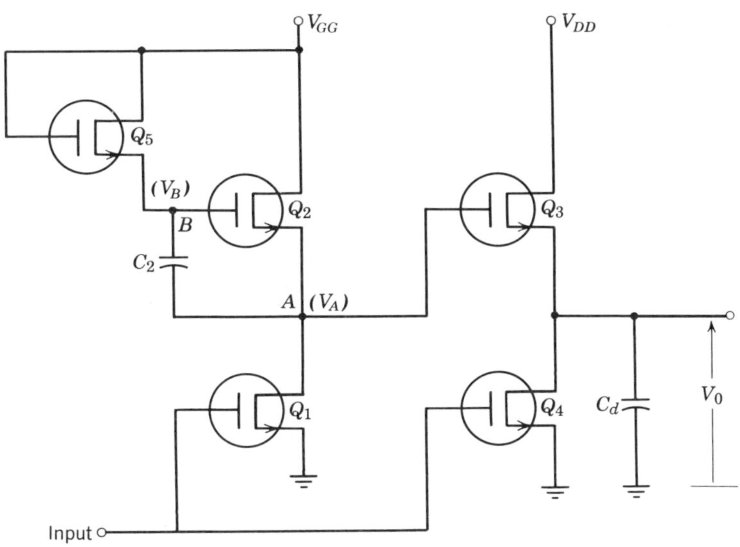

Figure 7.23. Bootstrap push-pull MOSFET driver circuit. (From *MOS Integrated Circuits*, edited by W. Penney and L. Lau, © 1972 by Litton Educational Publishing, Inc. Reprinted by permission of Van Nostrand Reinhold Company.)

V_{GG}. V_A causes Q_3 to be turned on, and this charges C_d until V_0 reaches its final value $V_{GG} - V_{T3}$ or V_{DD}, whichever is smaller. Because V_{GS2} is larger, the transient goes faster than in the previous case.

2. *Input switches from low to high.*

Q_1 and Q_4 are turned on immediately. This lowers V_A rapidly; at the same time V_0 discharges through Q_4. By proper dimensioning of the circuit, V_A decreases much more rapidly than V_0, so Q_3, which was off to begin with, stays off during the transient. If needed, the device Q_5 now recharges the capacitance C_2 until V_B reaches the value $V_{GG} - V_{T5}$. This transient goes about as fast as in the previous case.

7.9 COMPARISON OF LOGIC CIRCUITS*

Table I Comparison of the Major Integrated Circuit Digital Logic Families

Parameter	RTL	DTL	TTL	ECL	MOS	CMOS
Basic gate[a]	Nor	Nand	Nand	Or, Nor	Nor, Nand	Nor, Nand
Fan-out[b]	5	8	10	25	20	>50
Power dissipated per gate (mW)[c]	12	8–12	12–22	40–55	0.2–10	0.01 static, 1 at 1 MHz
Noise immunity	Nominal	Good	Very good	Good	Nominal	Very good
Propagation delay per gate (ns)[d]	12	30	12–6	4–1	300	70
Clock rate (MHz)[e]	8	12–30	15–60	60–400	2	5
Number of functions	High	Fairly high	Very high	High	Low	Low

[a] Positive logic.
[b] Worst-case number of inputs the gate can drive.
[c] Typical; affected by temperature and frequency of operation.
[d] Propagation delay is usually defined as $T_{PD} = \frac{1}{2}(T_{\text{on}} + T_{\text{off}})$.
[e] Maximum frequency at which flip-flops operate. The actual clock rate is from one-half to one-tenth the frequency listed.

Here T_{on} is defined as the time difference between the instant V_{in} passes the 0 threshold upward and the instant V_{out} passes the 1 threshold downward. Similarly, T_{off} is defined as the time difference between the instant V_{in} passes the 1 threshold downward and the instant V_{out} passes the 1 threshold upward (Fig. 7.24). T_{on} and T_{off} follow from Chapters 3 and 4.

*L. S. Garrett, "Integrated Circuit Digital Logic Families," Parts I, II, and III, *IEEE Spectrum* (October–December 1970).

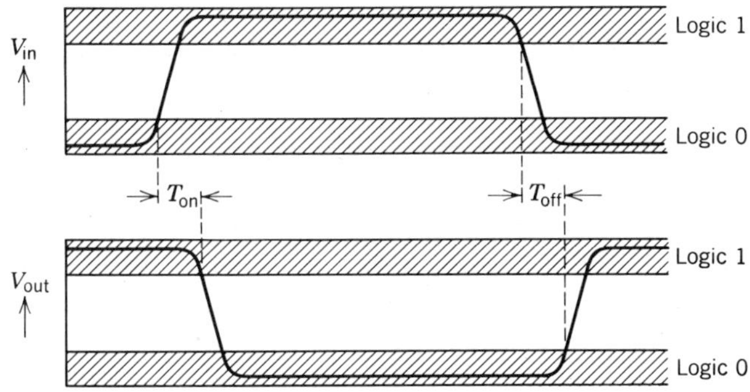

Figure 7.24 Definition of logical 0 and 1 of V_{in} and of V_{out}. Definition of T_{on} and T_{off}.

PROBLEMS

7.1 In Fig. 7.25 $V(t)=4.0$ V, $V_{BE0}=0.70$ V, $V_{CC}=3.0$ V, $\beta_F=50$, and Schottky barrier diode D clamps V_0 at 0.30 V when I_B is sufficiently large. Find I_B, I_C, I'_B, and I'_C.

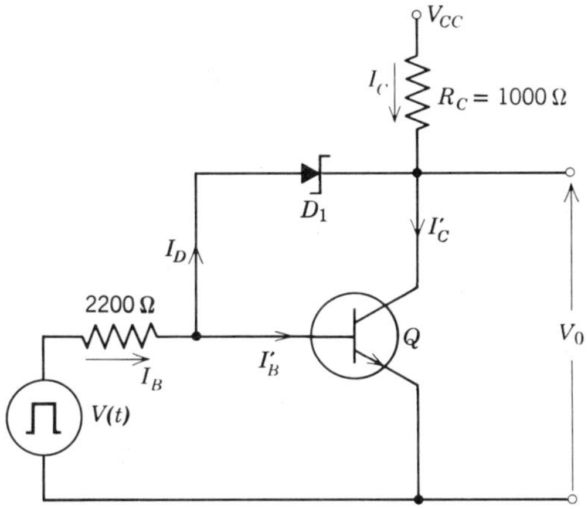

Figure 7.25

7.2 In Fig. 7.3b find the turn-on voltages of gates A and B if $V_{BE0}=0.70$ V and $V_{CE,\text{sat}}=0.20$ V.

7.3 In Fig. 7.4b find the turn-on voltages of gates A and B if $V_T=-3.0$ V and $V_{01}=V_{on}=-1.0$ V is the drain voltage of Q_1 when operating in the ohmic regime.

7.4 In Fig. 7.7, $V_{CC}=6.0$ V, $R_c=2000\ \Omega$, $R=2000\ \Omega$, $V_{BE0}=0.70$ V, $V_{CE,\text{sat}}=0.0$ V, and $\beta_F=30$. Find the voltage that must be supplied to input A to make Q_1 saturated if Q_2 and Q_3 are off.

7.5 In the circuit in Fig. 7.8 the transistor Q draws the largest current and has $V_{BE0}=0.70$ V. If $R_c=2000\ \Omega$, find the value of V_{CC} for which transistor Q_2 is just saturated if $I_{B2}=0.10\ I_{B1}$, $\beta_F=20$, and $V_{CE,\text{sat}}=0.0$ V.

7.6 The transistors in Fig. 7.9 are made on the same chip and are identical. $V_{CE,\text{sat}}=0.12$ V. Calculate the ratio of the currents flowing to inputs B and A' if transistors Q_1' and Q_1 are saturated.

7.7 In Fig. 7.10*a*, $V_{CC}=6$ V, $R_c=2000\ \Omega$, $V_{CE,\text{sat}}=0.0$ V, and $\beta_F=30$.

(a) Find the base current I_{BS} for which the transistor is just saturated. Then take I_B three times as large to ensure saturation.
(b) If $R_b=1000\ \Omega$, find the current that must flow into input A to make $I_B=3I_{BS}$ if gates B and C are off.

7.8 (a) In the circuit in Fig. 7.11*b* find how many diodes must replace D_4 to make gate A turn off at 2.20 V. Each diode has a turn-on voltage of 0.70 V, and so has transistor Q.
(b) Diode D_4 is replaced by a zener diode having a breakdown voltage of 4.0 V. What is the turn-off voltage of gate A now? *Hint*: Assume that the diode being turned off has a voltage drop of 0.60 V at turn-off.

7.9 In the circuit in Fig. 7.11*c* the output Y feeds into n similar gates. $V_{CC}=4.0$ V, $R_c=1000\ \Omega$, $R_1=3000\ \Omega$, $V_{CE,\text{sat}}=0.0$ V, $\beta_F=20$, and $I_B=3I_{BS}$, where I_{BS} is the base current needed to drive the transistor to saturation. What is the fan-out capability under worst-case conditions?

7.10 The TTL gate in Fig. 7.13 has $V_{CC}=4.9$ V, $R_b=1750\ \Omega$, $R_c=1000\ \Omega$, and $C_0=20$ pF; Q_1 has $\beta_F=100$, $V_{BE0}=0.70$ V, and $V_{CE,\text{sat}}\cong 0.00$ V.

(a) Find the time it takes to turn Q_1 on by means of T_1, after it has been off.
(b) Find the output current capability of Q_1 when on.
(c) Find the time it takes to turn Q_1 off by means of T_1, after it has been on.

7.11 In the direct-drive totem-pole circuit in Fig. 7.14 $R_1=R_2=2000\ \Omega$, $V_1=4.0$ V, $V_{BE0}=0.70$ V, $V_{CE,\text{sat}}=0.0$ V, $\beta_F=50$, $V_{CC}=5.0$ V, and $C_0=50$ pF.

(a) If Q_1 and Q_2 are identical transistors, how much current I_2 can be forced into Q_2 so that the transistor remains saturated.
(b) If Q_2 is off, find the maximum current that can be drawn out of Q_2 if the minimum value needed to make V_0 a 1 output is 2.0 V.
(c) Find the turn-off time of the output if V_0 was at 3.0 V.
(d) Find the turn-on time of the output.

7.12 In the totem-pole circuit in Fig. 7.15, $R_2=1500\ \Omega$, $R_1=700\ \Omega$, $I_{B1}=2.0$ mA, and $C_0=50$ pF. All transistors and diodes have a turn-on voltage of 0.70 V, $V_{CE,\text{sat}}\cong 0.00$ V, and the transistors have $\beta_F=100$. If $V_{CC}=4.9$ V, find

(a) The current I_{B2} when Q_1 is on and saturated.
(b) The current capability of Q_2 in (a).
(c) The turn-off time of the output of Q_2 if $V_0(t)$ started at 2.0 V.
(d) The current capability of Q_3 when Q_1 is off, if $V_0(t)$ is not allowed to drop below 2.0 V.
(e) The turn-on time of the output due to Q_3.

7.13 In the circuit in Fig. 7.19 $R_1=R_2=300\ \Omega$, $V_{BE0}=0.70$ V, $R_E=R_0=1000\ \Omega$, $V_{EE}=-4.9$ V, and $V_{\text{ref}}=-1.4$ V.

(a) For what values of V_A and V_B are A and B 0's?
(b) What are the values of V_4 and V_5 in (a)?
(c) For what values of V_A and V_B are A and B 1's?
(d) What are the values of V_4 and V_5 in (c)?

7.14 Explain the TTL circuit in Fig. 7.26. $V_{BE0}=0.70$ V, $V_{CE,\text{sat}}=0.20$ V, where applicable, $\beta_F=100$ for all transistors, and $C_0=100$ pF. Explain especially the current capabilities for a 0 output and a 1 output. Assume for a 1 output that $V_y=3.0$ V. Find also the turn-on time (time needed for 90% completion of the transient going from 0.20 V to 3.0 V) and the turn-off time (full completion of the transient going from 3.0 V to 0.20 V) for the output of Q_4.

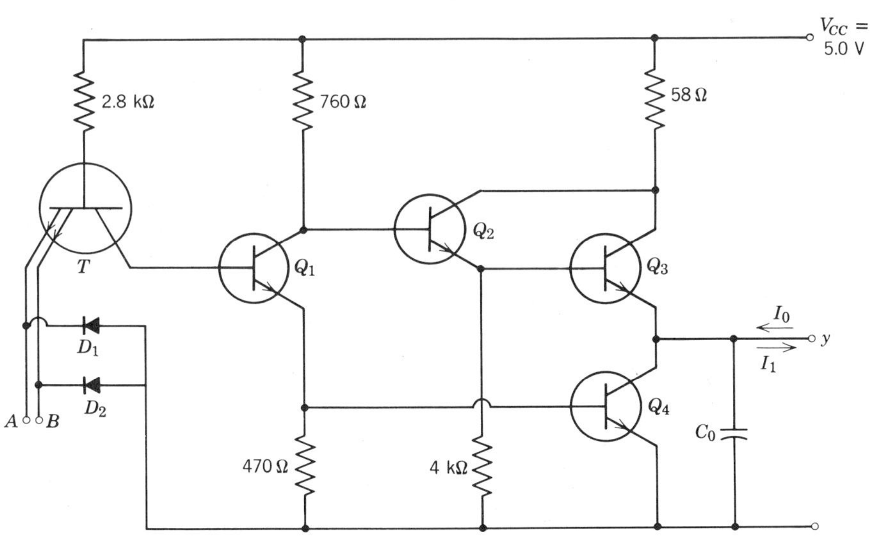

Figure 7.26

7.15 Discuss why the circuit in Fig. 7.27 is so much slower than the circuit in Fig. 7.26.

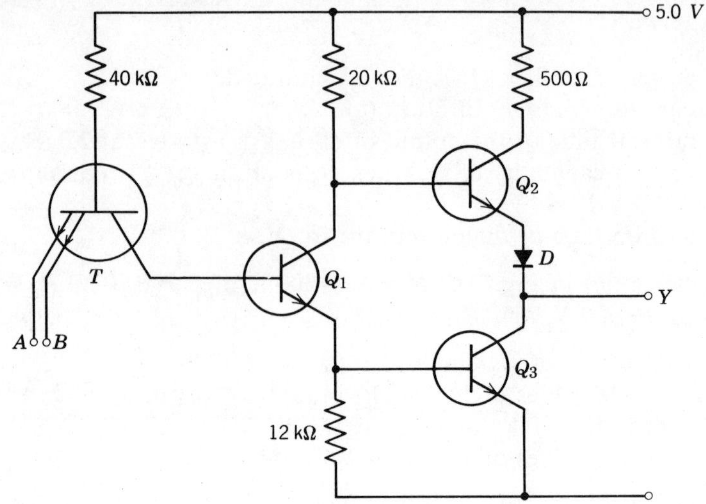

Figure 7.27

7.16 Explain the operation of the two-input nor gate in Fig. 7.28.

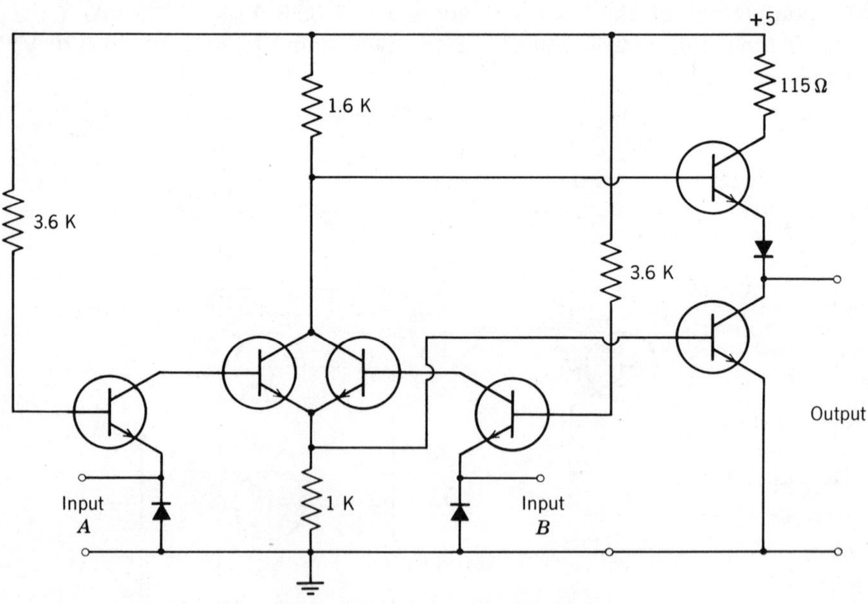

Figure 7.28. Two-input nor gate.

CHAPTER

8 Time Bases and Blocking Oscillators

This chapter deals with time bases (Section 8.1) and blocking oscillators (Section 8.2). Time bases are circuits that generate a linear voltage or current waveform for the interval $-T/2<t<T/2$, that is periodically repeated outside this interval. Section 8.1 indicates how this can be done. Blocking oscillators are either of the monostable (generation of a single waveform when triggered) or astable (generation of a periodic waveform without triggering) kind; they use a pulse transformer as the circuit element. Section 8.2 discusses these circuits in some detail.

8.1 TIME BASES

It is often necessary to make a sawtoothlike voltage waveform

$$V(t)=2V_1\frac{t}{T} \qquad \text{for } -\frac{T}{2}<t<\frac{T}{2} \tag{8.1}$$

and periodically repeated outside this interval, or a sawtoothlike current waveform

$$I(t)=2I_1\frac{t}{T} \qquad \text{for } -\frac{T}{2}<t<\frac{T}{2} \tag{8.2}$$

and periodically repeated outside this interval. Such voltages and currents are called *sweep* voltages and currents, because they are mainly used to sweep the light spot of a cathode ray tube over a screen. The circuits producing these voltage and current waveforms are called *sweep circuits* or *time bases*. Here we discuss the basic features of such circuits.

8.1a Fundamentals of Voltage and Current Sweeps

Figure 8.1*a* shows a constant-current generator I_0 applied to a capacitor C which can be short-circuited by a switch S when the need arises. Let the switch S be opened at $t=0$; then the output voltage $V_0(t)$ changes linearly with time as

$$V_0(t)=\frac{I_0 t}{C} \tag{8.3}$$

At $t=T$ let the switch S be closed for a short time; then $V_0(t)$ becomes zero. If the switch is then reopened for another time interval T, the waveform repeats itself.

We see that the peak to peak amplitude of the sweep voltage is given as

$$V_1=\frac{I_0 T}{C} \tag{8.3a}$$

Usually T and V_1 have required values; Eq. (8.3a) tells us what value of the capacitance C must be used for a given current I_0.

Figure 8.1*b* shows an emf V_0 applied to an inductor L by means of a switch S. Let the switch S be closed at $t=0$; then, for $R=\infty$

$$L\frac{dI}{dt}=V_0 \qquad \text{or} \qquad I(t)=\frac{V_0 t}{L} \tag{8.4}$$

so the current $I(t)$ flowing through the inductor varies linearly with time.

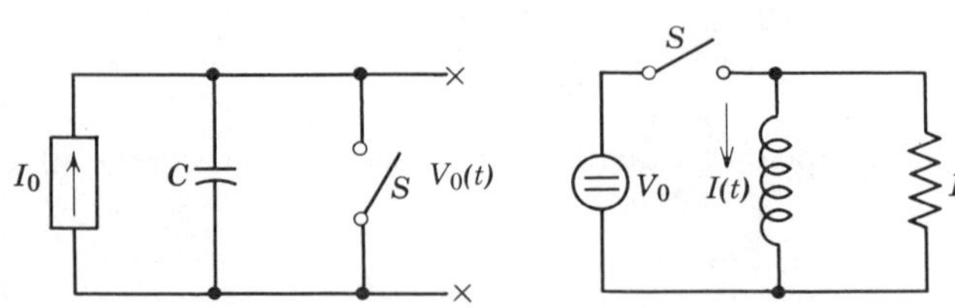

Figure 8.1. (*a*) Simple circuit producing a voltage sweep. (*b*) Simple circuit producing a current sweep.

At $t = T$ let the switch S be opened; then $I_0(t)$ becomes zero. If we ignore for a moment the infinite voltage developed across the terminals of the coil, lasting for an infinitely short time, we can then close the switch S and let the waveform repeat itself.

We see that the peak to peak amplitude of the sweep current is given as

$$I_1 = \frac{V_0 T}{L} \tag{8.4a}$$

Usually T and I_1 have required values; Eq. (8.4a) tells us what value V_0 of voltage must be used for a given inductance L.

We now investigate what must be done to reduce the voltage developed across the inductor L. Let a resistor R be connected in parallel with L. This does not alter Eqs. (8.4) and (8.4a). If at $t = T$ the switch S is opened the current I_1 flows through R at $t = T$ (for $t > T$), this circulating current decays as

$$I(t) = I_1 \exp\left[-\frac{R}{L}(t - T)\right] \tag{8.5}$$

The voltage developed across L is therefore

$$V_1 = L\frac{dI}{dt} = -I_1 \frac{R}{L} \exp\left[-\frac{R}{L}(t - T)\right] \tag{8.5a}$$

so the initial voltage amplitude is $-V_0 TR/L^2$. By proper choice of R this can be made to stay within tolerable limits. The time constant τ of the circuit is L/R.

8.1b Voltage Sweep Circuits

Figure 8.2*a* shows a linear voltage sweep circuit in which the switch S has been replaced by a pulsed transistor Q. The constant-current generator is best made by connecting the gate of a JFET to the source. In this case $I_0 = I_{DSS} = KV_p^2$ as long as $V_{DS} \geqslant -V_p$, where V_p is the pinch-off voltage of the JFET.*

We assume that the transistor Q is saturated for $t < 0$ and that it is turned off at $t = 0$. Then $V_0(0) = V_{CE,\text{sat}} \cong 0$. For $t > 0$ we thus have

$$V_0(t) = \frac{I_0 t}{C} \tag{8.6}$$

*Two-terminal JFET-like devices are made commerically, for which $\partial I_D / \partial V_{DS}$ is small, so the device is practically ideal.

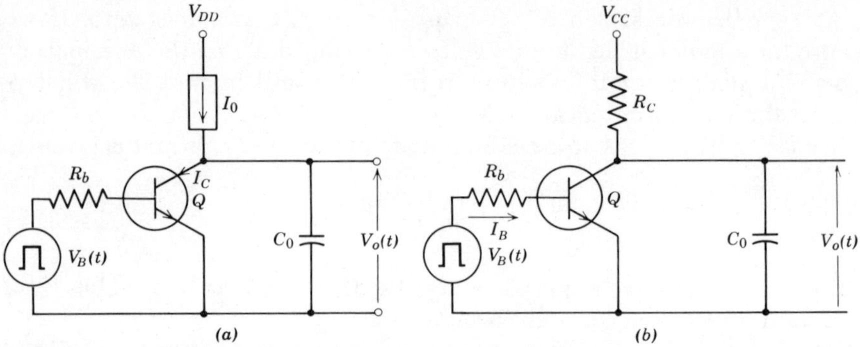

Figure 8.2. (*a*) Circuit producing a linear voltage sweep. (*b*) Circuit producing a voltage sweep that is not so linear.

For a JFET current generator this linear transient becomes nonlinear when $V_0(t) = V_{DD} + V_p$. To prevent this from happening we turn the transistor on before $V_0(t)$ reaches this value. This means that T must be chosen so that

$$T \leqslant \frac{C(V_{DD} + V_p)}{I_0} \tag{8.6a}$$

Usually, however, T is given and Eq. (8.6a) determines C or $(V_{DD} + V_p)$.

For $t > T$ the transistor is turned on for a short time to discharge the capacitor C. Let $V_B(t) = V_1$ for $t > T$; then $I_B = (V_1 - 0.70)/R_b$ and $I_C = \beta_F I_B = \beta_F(V_1 - 0.70)/R_b$. By proper choice of V_1 and R_b one can make I_C so large that $I_0 \ll I_C$. In this case the capacitor C is discharged during the time t_1, where

$$t_1 = \frac{CV_0(T)}{I_C} \tag{8.7}$$

Now, according to Eq. (8.6),

$$T = \frac{CV_0(T)}{I_0} \tag{8.7a}$$

so

$$t_1 = T\frac{I_0}{I_C} \tag{8.7b}$$

At $t = T + t_1$, $V_0(t) = V_{CE,\text{sat}} \cong 0$; the transistor Q is now turned off and the cycle repeats itself.

It is not difficult to make t_1 small in comparison with T by proper choice of I_C.

If one does not require absolute linearity, the simple circuit in Fig. 8.2*b* can be used. Here the current generator I_0 has been replaced by a resistor R_c, and the supply voltage is denoted by V_{CC}.

Suppose at $t=0$ the transistor is turned off, and $V_0(0)=V_{CE,\text{sat}}\cong 0$. The output voltage $V_0(t)$ then varies as

$$V_0(t)=V_{CC}\left[1-\exp\left(-\frac{t}{R_cC}\right)\right] \tag{8.8}$$

For $t\ll R_cC$ this may be approximated as

$$V_0(t)=V_{CC}\frac{t}{R_cC} \tag{8.8a}$$

which varies linearly with time. But for larger t we must use more terms in the Taylor expansion of $V_0(t)$ with respect to t

$$V_0(t)=V_{CC}\frac{t}{R_cC}\left[1-\frac{1}{2}\frac{t}{R_cC}+\cdots\right] \tag{8.8b}$$

so the percentage deviation d from linearity at the end ($t=T$) of the sweep is approximately

$$d=\frac{50T}{R_cC} \tag{8.8c}$$

It depends on the application how much deviation from linearity can be tolerated. For television displays the requirements are not too stringent.

In order to obtain a large sweep-voltage amplitude V_1 one must make V_{CC} quite large, and one should use high-voltage transistors. Substituting $t=T$ in Eq. (8.8a) yields

$$V_1\cong V_{CC}\frac{T}{R_cC}=V_{CC}\frac{d}{50} \tag{8.9}$$

8.1c Current Sweep Circuits

Figure 8.3*a* shows a current sweep with a transistor Q as the switching element. The transistor is normally on and saturated but can be turned off by a negative voltage pulse applied through two diodes (D_1 and D_2) in

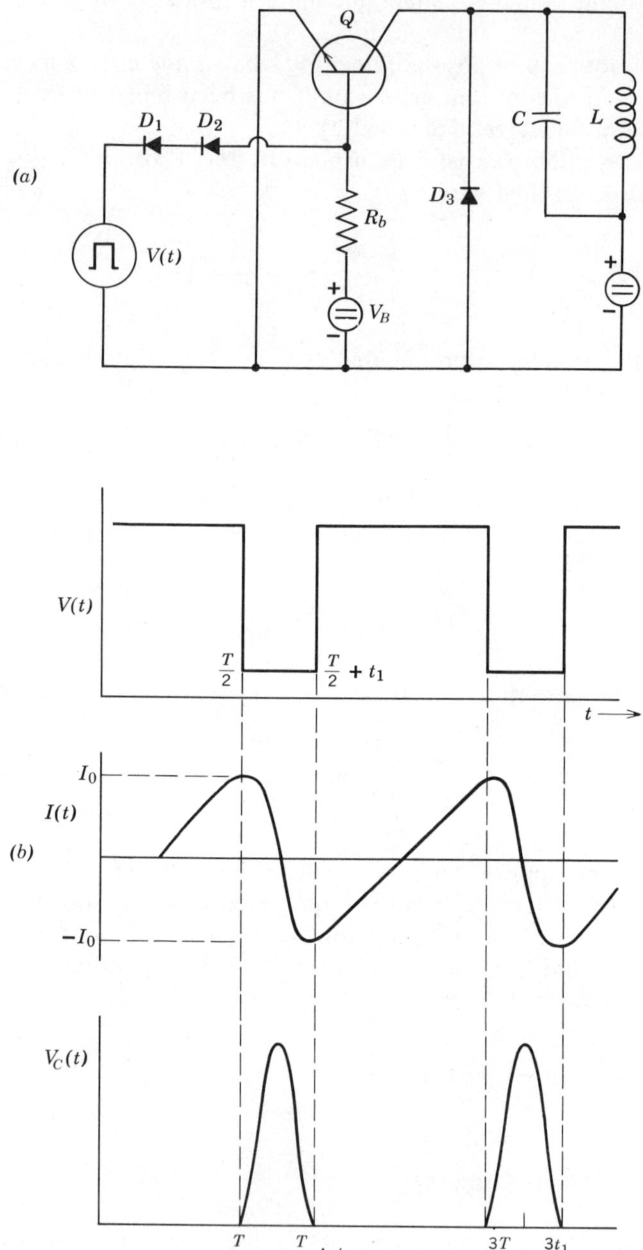

Figure 8.3. (*a*) Circuit producing a linear current sweep with flyback capability. (*b*) Waveform of $V(t)$, $I(t)$, and $V_C(t)$.

series. The aim of the circuit is to produce a sawtooth current through the inductor L, where

$$I(t)=2I_0\frac{t}{T} \qquad \text{for } -\frac{T}{2}<t<\frac{T}{2} \tag{8.10}$$

and periodically repeated outside this interval. The circuit also contains a diode (D_3) between the collector and the ground, which is normally back-biased, and a small capacitor C, the use of which will soon become clear.

At $t=0$ let the transistor Q be on and saturated and let $I(t)=0$. Then $I(t)$ increases linearly with t for $t>0$. If $I(t)$ reaches the value I_0 at $t=T/2$, a negative voltage pulse is applied to the base, which turns the transistor Q off. This sets up oscillations in the LC circuit so that

$$I(t)=I_0\cos\omega_1\left(t-\frac{T_1}{2}\right) \qquad \text{where } \omega_1=\frac{1}{(LC)^{1/2}} \tag{8.11}$$

is the resonance frequency of the LC circuit. For $\omega_1(t-T_1/2)=\pi/2$ the collector voltage $V_C(t)$ of Q passes through a maximum, and $I(t)=0$; for $\omega_1(t-T_1/2)=\pi$ the collector voltage $V_C(t)$ just turns negative, and $I(t)=-I_0$. At this moment diode D_3 becomes forward-biased, and the oscillation stops because the current I_0 flows through D_3 instead of C.

If at the same time the negative voltage pulse on the base of Q is turned off, the current through L increases linearly from $-I_0$ to $+I_0$. At the instant when $I(t)=+I_0$ the process is repeated, so a waveform [Eq. (8.9)] is generated (Fig. 8.3*b*).

It should be noted that the change in current from $+I_0$ to $-I_0$ is accomplished in a time interval t_1, such that

$$\omega_1 t_1=\pi \qquad \text{or} \qquad t_1=\pi(LC)^{1/2} \tag{8.12}$$

By proper choice of C this can be kept small. Figure 8.3*b* gives $V_C(t)$ as a function of time. It shows a very sharp peak at $T/2+t_1/2$. The transistor Q and the diode D_3 must therefore be high-voltage devices which have a breakdown voltage larger than this peak value.

In cathode ray tubes with magnetic deflection, the deflection coils producing the deflecting magnetic field serve as the inductor L in Fig. 8.3*a*.

8.2 THE BLOCKING OSCILLATOR

Blocking oscillators are of two types:

1. *Monostable blocking oscillators.*
These are normally off through proper bias of the device used in the circuit. After being triggered by a pulse, the circuit generates one complete waveform of duration t_p and then stays in the stable condition unless triggered by another pulse.

2. *Astable blocking oscillators.*
This circuit is biased opposite the one in the previous case, so it has no stable operating point. The circuit generates a periodic waveform in which the device is alternately on for a time t_p and off for a time t_{off}.

Blocking oscillators use pulse transformers as the feedback element. In order to understand the operation of the pulse transformer we must bear in mind its equivalent circuit. According to Appendix 1, a pulse transformer can be represented by an ideal transformer of turns ratio n with a small spreading inductance L_s in series and a much larger magnetizing inductance L_m in parallel. Figure 8.4 shows the equivalent circuit; the relatively small series resistance of the transformer is not shown.

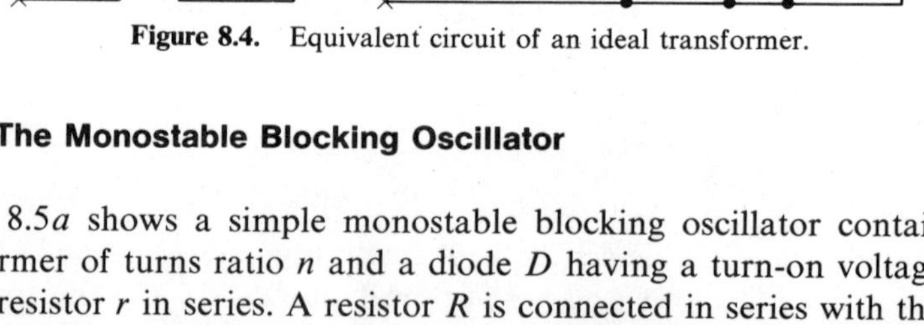

Figure 8.4. Equivalent circuit of an ideal transformer.

8.2a The Monostable Blocking Oscillator

Figure 8.5*a* shows a simple monostable blocking oscillator containing a transformer of turns ratio n and a diode D having a turn-on voltage V_{BE0} with a resistor r in series. A resistor R is connected in series with the base. Typically R is a few thousand ohms, r is of the order of 10 Ω, and $n \cong \frac{1}{2}$. The voltage V_{BB} is zero for the monostable blocking oscillator, so the transistor is normally off.

Suppose at $t=0$ the collector is triggered, and its voltage is reduced to zero for a short time. If spreading inductance effects are neglected the

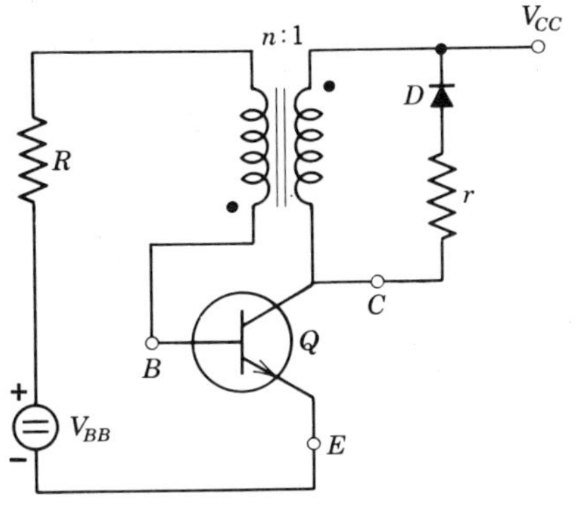

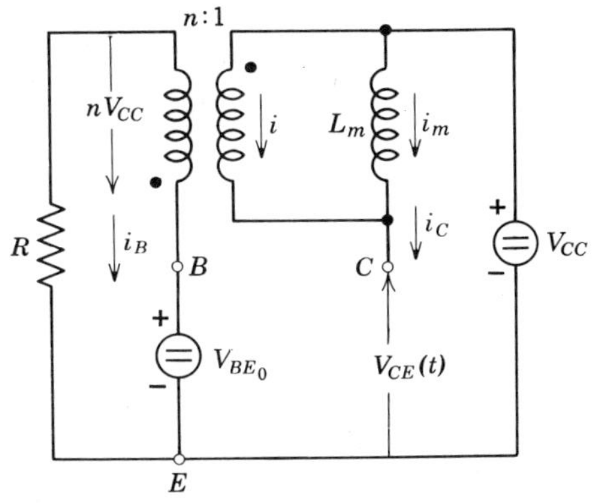

Figure 8.5. (*a*) Monostable blocking oscillator circuit. (*b*) Equivalent circuit.

voltage V_{CC} appears practically instantaneously across the transformer, and on the other side of the transformer the voltage is nV_{CC} with the polarity as shown. The base current is therefore

$$i_B = \frac{nV_{CC} - V_{BE0}}{R} \tag{8.13}$$

where V_{BE0} is the turn-on voltage of the transistor. The current flowing through the collector side of the ideal transformer is then

$$i = ni_B = n\frac{nV_{CC} - V_{BE0}}{R} \tag{8.14}$$

and the magnetizing current i_m flowing through L_m is

$$i_m = \frac{V_{CC}t}{L_m} \tag{8.15}$$

since $V_{CE,\text{sat}} \cong 0$, so the collector current

$$i_C(t) = i + i_m = \frac{n(nV_{CC} - V_{BE0})}{R} + \frac{V_{CC}t}{L_m} \tag{8.16}$$

At first the transistor is saturated but, if β_F is the current amplification factor of the transistor, there comes a moment when

$$i_C(t) = \beta_F i_B \qquad \text{or} \qquad \frac{n(nV_{CC} - V_{BE0})}{R} + \frac{V_{CC}t}{L_m} = \beta_F\frac{nV_{CC} - V_{BE0}}{R} \tag{8.17}$$

At this moment the transistor comes out of saturation and, since the loop gain is larger than unity, the transistor is quickly driven to cutoff by regeneration. The duration t_p of the pulse is obtained by solving t; this yields

$$t = t_p \cong \beta_F \frac{L_m}{R}\left(\frac{nV_{CC} - V_{BE0}}{V_{CC}}\right) \tag{8.17a}$$

since $\beta_F \gg n$.

At this point a magnetizing current

$$I_0 = \frac{V_{CC} t_p}{L_m} = \beta_F \frac{nV_{CC} - V_{BE0}}{R} \tag{8.18}$$

flows through L_m; it must return through the diode. Therefore $V_{CE}(t)$, which was practically zero during the pulse, becomes

$$V_{CE}(t_p^+) = V_{CC} + V_{BE0} + I_0 r \tag{8.19}$$

The current $I(t)$ flowing through L_m decays as

$$I(t) = I_0 \exp\left[-\frac{r}{L_m}(t - t_p)\right] \tag{8.20}$$

and more rapidly when the diode series resistance becomes significant. The recovery time is therefore a few times the time constant L_m/r.

Figure 8.6 shows the waveform of $i_B(t)$, $i_C(t)$, and $V_{CE}(t)$. It goes without saying that the best results are obtained if β_F is not too large. For this reason the transistor is often used in a common base connection.

The pulse duration t_p depends on β_F. The circuit is therefore not very practical, since β_F depends on temperature and differs widely from transistor to transistor. However, it shows the basic features of the monostable blocking oscillator, and this is all that is needed at this stage of the discussion. If the "bump" in $V_{CE}(t)$ is undesirable, it can be eliminated by appropriate waveshaping techniques.

Example 1. If $V_{CC} = 10$ V, $n = \frac{1}{2}$, $R = 2000\ \Omega$, $L_m = 5$ mH, $r = 20\ \Omega$, $\beta_F = 50$, and $V_{BE0} = 0.70$ V, find i_B during the pulse, the pulse duration t_p, the maximum magnetization current I_0, the initial output voltage after turn-off, and the recovery time.

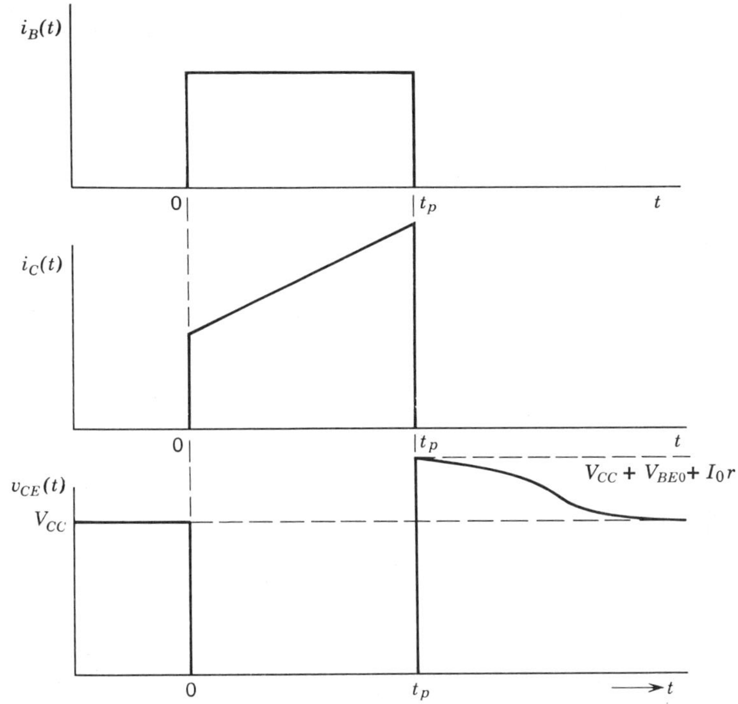

Figure 8.6. Waveform of $i_B(t)$, $i_C(t)$, and $V_{CE}(t)$ for the circuit in Fig. 8.5.

Answer. We neglect V_{BE0}, except in $V_{CE}(t_p^{+})$. Then

$$i_B = \frac{nV_{CC}}{R} = \frac{\frac{1}{2}\times 10}{2000} = 2.5 \text{ mA}$$

$$t_p = \beta_F \frac{nL_m}{R} = 50\times \frac{\frac{1}{2}\times 5\times 10^{-3}}{2000} = 0.625\times 10^{-4} \text{ s}$$

$$I_0 = \frac{V_{CC}t_p}{L_m} = \frac{10\times 0.625\times 10^{-4}}{5\times 10^{-3}} = 0.125 \text{ A}$$

$$V_{CE}\left(t_p^{+}\right) = 10 + 0.70 + 2.5 = 13.2 \text{ V}$$

$$\text{Recovery time} \cong 4\frac{L_m}{r} = \frac{4\times 5\times 10^{-3}}{20} = 10^{-3} \text{ s}$$

8.2b The Astable Blocking Oscillator

The circuit used in Fig. 8.5 becomes an astable blocking oscillator when V_{BB} is made positive, for example, equal to a few volts.

Suppose at $t=0$ the voltage V_{BE} between the base and the emitter reaches 0.70 V; then the transistor is turned on. Soon the loop gain of the circuit is larger than unity, and the transistor currents rise so rapidly by regenerative action that the device becomes saturated. This is practically instantaneous.

Now a pulse is generated as before. In analogy with the previous case

$$i_B(t)=\frac{nV_{CC}+V_{BB}-V_{BE0}}{R}=i_B \tag{8.21}$$

$$i_C=ni_B+\frac{V_{CC}t}{L_m}=\frac{n(nV_{CC}+V_{BB}-V_{BE0})}{R}+\frac{V_{CC}t}{L_m}=\beta_F i_B \tag{8.22}$$

The duration t_p of the pulse is, since $\beta_F \gg n$,

$$t_p \cong \frac{\beta_F L_m}{V_{CC}} i_B=\beta_F \frac{L_m}{R}\frac{nV_{CC}+V_{BB}-0.70}{V_{CC}} \tag{8.23}$$

The maximum magnetizing current is again

$$I_0=\frac{V_{CC}t_p}{L_m} \tag{8.24}$$

Finally, for $t>t_p$,

$$V_{CE}(t)=V_{CC}+V_{BE0}+I_0 r\exp\left[-\frac{r}{L_m}(t-t_p)\right] \tag{8.25}$$

so

$$V_{BE}(t)=V_{BB}-n\left\{V_{BE0}+I_0 r\exp\left[-\frac{r}{L_m}(t-t_p)\right]\right\} \tag{8.26}$$

As soon as V_{BE} reaches the value V_{BE0}, the transistor is turned on and the cycle repeats itself. Substituting $v_{BE}(t)=V_{BE0}$ and solving for $t-t_p$ yields, for the off time of the circuit,*

$$t - t_p = t_{\text{off}} = \frac{L_m}{r} \ln\left[\frac{nI_0 r}{V_{BB} - (n+1)V_{BE0}}\right] \qquad (8.27)$$

Example 2. If $V_{CC} = 10$ V, $n = \frac{1}{2}$, $R = 2000\ \Omega$, $L_m = 5$ mH, $r = 20\ \Omega$, $\beta_F = 50$, $V_{BE0} = 0.70$ V, and $V_{BB} = 2.0$ V, find t_p and t_{off}.

Answer.

$$t_p = 50 \times \frac{5 \times 10^{-3}}{2000}\left(\frac{5.00 + 2.00 - 0.70}{10}\right) \cong 0.8 \times 10^{-4}\ \text{s}$$

$$I_0 \cong \frac{10 \times 0.8 \times 10^{-4}}{5 \times 10^{-3}} = 160\ \text{mA}$$

$$t_{\text{off}} = \frac{5 \times 10^{-3}}{20} \ln\left(\frac{1.60}{2.00 - 1.05}\right) \cong 1.3 \times 10^{-4}\ \text{s}$$

In the circuit discussed here t_p, I_0, and t_{off} depend on β_F. Circuits can be designed that operate practically independently of transistor parameters and have a lower duty cycle. It is beyond the scope of this book to consider them in detail.

PROBLEMS

8.1 (a) In the circuit in Fig. 8.2*a* find C such that $V_0(t)$ goes from 0 V to 10 V in 10^{-4} s if $I_0 = 4$ mA.
(b) How large must R_b be if the voltage $V_0(t)$ returns to 0 V in 10^{-6} s if $\beta_F = 100$ and $V_1 = 5.0$ V.

8.2 In the circuit in Fig. 8.2*b*, $V_{CC} = 100$ V.

(a) Find the peak output voltage if $d = 5\%$.
(b) Find C if $R_c = 2000\ \Omega$ and $T = 10^{-4}$ s.

*As Eq. (8.27) indicates, the circuit must satisfy the condition $nI_0 r > V_{BB} - (n+1)V_{BE0}$, otherwise the circuit restarts immediately.

8.3 (a) In the current sweep circuit in Fig. 8.3a, $L = 1.0$ mH. Find the value of C such that $t_1 = 10^{-6}$ s.
(b) If $V_{CC} = 20$ V and $T = 10^{-4}$ s, find I_0.

8.4 (a) Develop the theory of the circuit in Fig. 8.7 in analogy with the circuit in Fig. 8.5. The collector is triggered at $t = 0$.
(b) If $R_e = 50\ \Omega$, $L_m = 10^{-2}$ H, $V_{CC} = 10$ V, and $V_{BE0} = 0.70$ V, find the duration of the pulse if the common base current amplification factor is 0.98 and $n = \frac{1}{2}$.

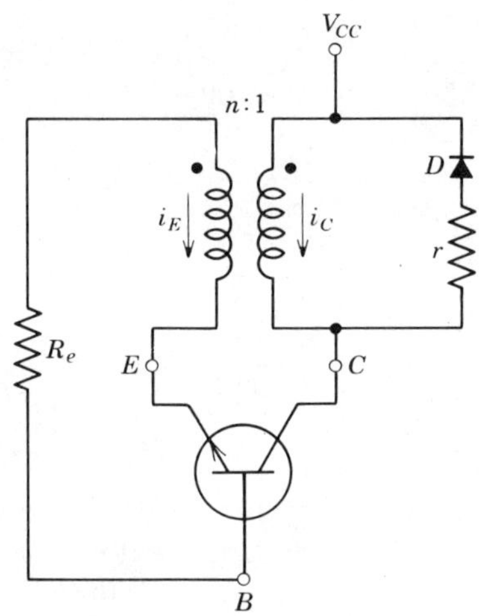

Figure 8.7. Astable blocking oscillator.

CHAPTER

9 Oscillators

Oscillators are usually made by feeding so much signal back from the output of an amplifier to the input that the output voltage can be sustained (Section 9.1). Oscillators can also be made by connecting a negative conductance in parallel with a parallel tuned circuit, or, a negative resistance in series with a series tuned circuit (Section 9.2). We show that the two methods of approach are identical. Section 9.3 discusses a laser as an example of a "distributed" oscillator.

9.1 OSCILLATORS AND FEEDBACK

9.1a The Nyquist and the Barkhausen Conditions

Figure 9.1 shows an amplifier having an amplification μ, with a feedback network feeding back a voltage βv_0 to the input. We now investigate how the output voltage v_0 is modified by the feedback.

Let v_s be the voltage applied to the feedback amplifier; then the input voltage v_i is

$$v_i = v_s + \beta v_0 \tag{9.1}$$

But

$$v_0 = \mu v_i \tag{9.2}$$

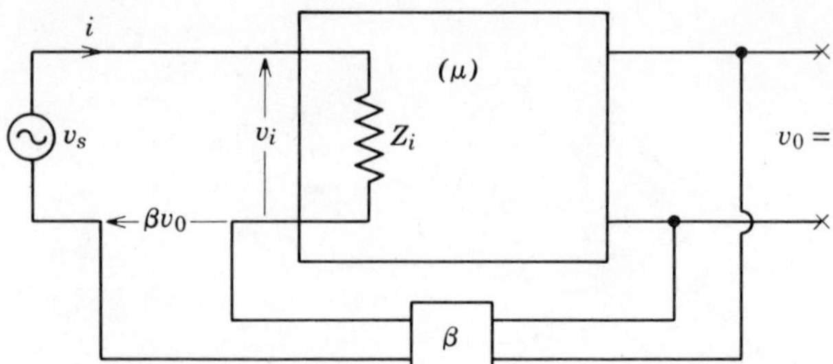

Figure 9.1. Feedback amplifier of voltage amplification μ and voltage feedback factor β.

so

$$v_i = v_s + \mu\beta v_i \qquad \text{or} \qquad v_i = \frac{v_s}{1-\mu\beta} \tag{9.3}$$

hence

$$v_0 = \mu v_i = \frac{\mu}{1-\mu\beta} v_s \tag{9.4}$$

If $\mu\beta$ is real and positive and $\mu\beta > 1$, oscillations usually occur. The general conditions for stability and instability are quite complicated, but the following condition, known as the *Nyquist condition*, applies in most simple cases. *If the factor $\mu\beta$ is plotted in a complex plane with the frequency ω as a parameter, one obtains a closed curve; the circuit is stable if the curve does not encircle the point* (1,*j*0), *and is unstable if the curve encircles the point* (1,*j*0) *once*. The plot of $\mu\beta$ in the complex plane is called the *Nyquist plot*. Figure 9.2*a* shows the Nyquist plot for a stable circuit, and Fig. 9.2*b* the Nyquist plot for an unstable one.

We now investigate the input impedance Z_{in} "seen" by the signal source v_s in Fig. 9.1. If Z_i is the input impedance of the amplifier, the current i flowing into the circuit is

$$i_i = \frac{v_s + \beta v_0}{Z_i} = \frac{v_s}{Z_i}\left(1 + \frac{\mu\beta}{1-\mu\beta}\right) = \frac{v_s}{Z_i(1-\mu\beta)} \tag{9.5}$$

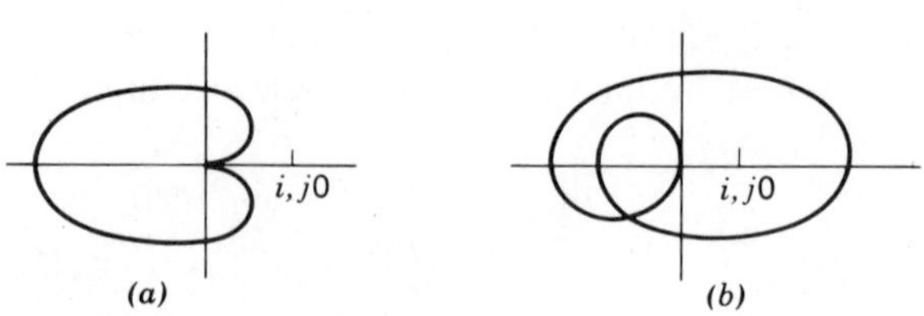

Figure 9.2. Examples of Nyquist plots. (*a*) Stable feedback circuit. (*b*) Unstable feedback circuit.

as is found by substituting for v_0. We have assumed that the signal βv_0 is developed across an impedance that is small in comparison with Z_i. The input impedance Z_{in} "seen" by the signal v_s is therefore

$$Z_{in} = \frac{v_s}{i_1} = Z_i(1 - \mu\beta) \tag{9.6}$$

The input impedance is thus zero for $\mu\beta = 1$. This important result is in agreement with results to be developed in Section 9.2.

If $v_s = 0$ and there is a frequency ω for which $\mu\beta$ is real and positive, there are three possibilities:

1. $\mu\beta < 1$. If oscillations are present at $t = 0$, their amplitude will decay exponentially.

2. $\mu\beta > 1$. If oscillations are present at $t = 0$, their amplitude will increase exponentially.

3. $\mu\beta = 1$. Oscillations, once present, will be maintained with a stable amplitude.

Now the amplification μ at a particular frequency ω for any amplifier depends on the input amplitude v_i. One generally finds for sufficiently large values of v_i that μ decreases with increasing amplitude v_i. Therefore, if we have a feedback amplifier with $\mu\beta > 1$ at a particular frequency ω for small amplitudes v_i, oscillations of frequency ω with growing amplitude will occur. However, $\mu\beta$ will ultimately decrease with increasing value of v_i and reach the value $\mu\beta = 1$. At this point the amplitude stops growing and a stable situation develops. We thus have the rule: *In an oscillator of frequency ω a stable amplitude will develop for which $\mu\beta = 1$.* This is called the *Nyquist condition* for *oscillators*.

A related condition, the *Barkhausen condition*, can be formulated as follows. *If in an oscillator the input lead is broken and an input voltage v_i is applied to the input, a voltage that is the same both in amplitude and in phase must appear at the other side of the break if stable oscillations are to occur.*

To show that this corresponds to the Nyquist condition, we turn to Fig. 9.3. If v_i is the voltage applied to the input, the output voltage $v_0 = \mu v_i$. If the part β of the output voltage is applied to the input, then, according to the Barkhausen condition,

$$v_i = \beta v_0 = \mu\beta v_i \qquad \text{or} \qquad \mu\beta = 1 \tag{9.7}$$

which is nothing but the Nyquist condition.

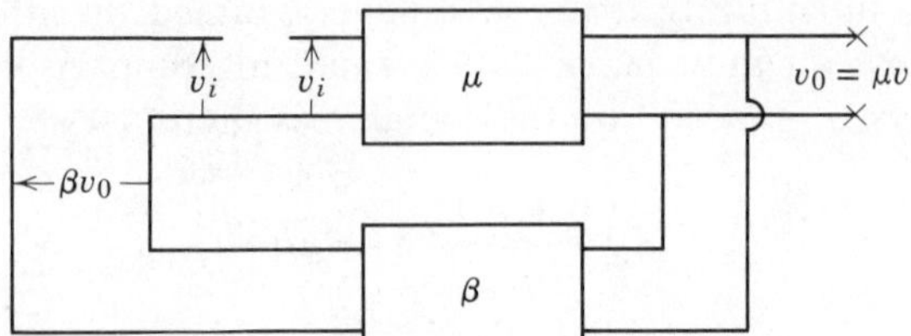

Figure 9.3. Illustration of the Barkhausen condition. In a feedback amplifier the input lead is broken and an input voltage v_i is applied; if the voltage returned on the other side of the break is equal to v_i in absolute value and phase, the restored circuit acts as an oscillator.

9.1b Applications of the Barkhausen Condition

As a first example we consider the Wien-bridge oscillator in Fig. 9.4*a*. It consists of a two-stage amplifier which is fed back to the input by means of an RC network with $R_1 = R_2 = R$ and $C_1 = C_2 = C$. The output resistance R_0 of the amplifier is small in comparison with R_1, the C_b's are blocking capacitors of negligible impedance, and the resistances R_g are so large that their effects can be neglected.

The Barkhausen condition can be illustrated with the help of Fig. 9.4*b*. Since the two impedances can be expressed as

$$R_1 + \frac{1}{j\omega C_1} = \frac{1 + j\omega C_1 R_1}{j\omega C_1} \tag{9.8}$$

and

$$\frac{R_2[1/(j\omega C_2)]}{R_2 + 1/(j\omega C_2)} = \frac{R_2}{1 + j\omega C_2 R_2} \tag{9.9}$$

the Barkhausen condition can be written as

$$\begin{aligned} v_i &= \mu v_i \frac{R_2/(1 + j\omega C_2 R_2)}{(1 + j\omega C_1 R_1)/(j\omega C_1) + R_2/(1 + j\omega C_2 R_2)} \\ &= \mu v_i \frac{j\omega C_1 R_2}{(1 + j\omega C_1 R_1)(1 + j\omega C_2 R_2) + j\omega C_1 R_2} \\ &= \mu v_i \frac{j\omega CR}{1 + 3j\omega CR - \omega^2 C^2 R^2} \end{aligned} \tag{9.10}$$

The first step is obtained by multiplying denominator and numerator by $j\omega C_1(1 + j\omega C_2 R_2)$, and the last step is obtained by putting $C_1 = C_2 = C$ and

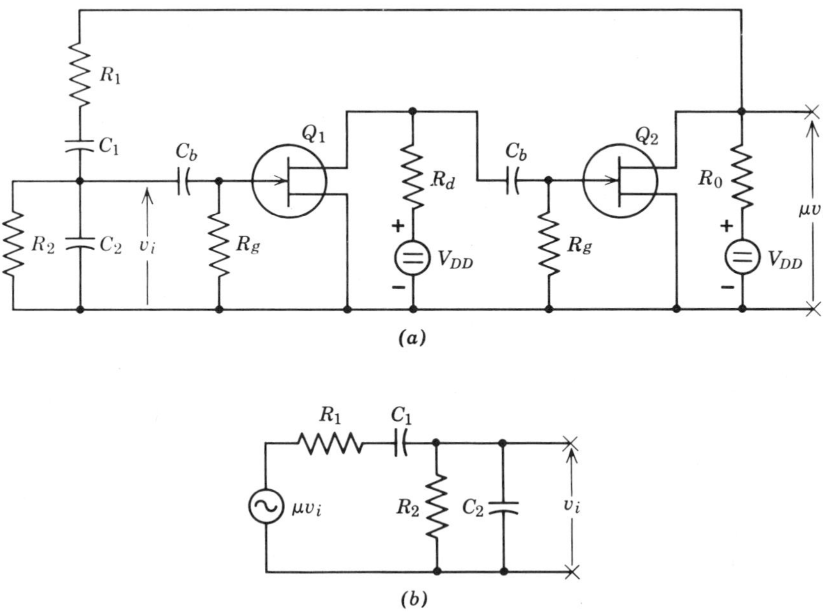

Figure 9.4. Wien-bridge oscillator. (*a*) Circuit. (*b*) Simplified equivalent circuit.

$R_1 = R_2 = R$. Dividing by v_i, and multiplying both sides by $1 + 3j\omega CR - \omega^2 C^2 R^2$ yields

$$1 + 3j\omega CR - \omega^2 C^2 R^2 = j\omega CR\mu \tag{9.10a}$$

or, since the real parts must be equal and the imaginary parts equal,

$$1 - \omega^2 C^2 R^2 = 0 \qquad \text{or} \qquad \omega CR = 1 \qquad \text{and} \qquad \mu = 3 \tag{9.10b}$$

The first condition thus yields the frequency of oscillation; this condition indicates that ω changes by a factor of 10 when C changes by a factor of 10. The second condition gives the amplitude of the oscillation. To start the oscillation, the low-amplitude value of μ must be somewhat larger than 3. The amplitude v_i will then increase until the stability condition $\mu(v_i) = 3$ is reached.

Example 1. If $R = 1000\ \Omega$, for what value of C is the frequency of oscillation equal to 1 MHz?

Answer.

$$2\pi fCR = 1 \qquad C = \frac{1}{2\pi fR} = \frac{1}{2\pi \times 10^6 \times 10^3} = \frac{10^{-9}}{2\pi} = 160\ \text{pF}$$

Example 2. If $\mu = 6/(1+0.25v_i^2)$, find the value of v_i for which a stable amplitude is obtained.

Answer.

$$3 = \frac{6}{1+0.25v_i^2} \qquad 1+0.25v_i^2 = 2 \qquad 0.25v_i^2 = 1 \qquad v_i = 2\text{ V}$$

As a second example we consider the circuit in Fig. 9.5*a*. This can be discussed with the help of the Barkhausen condition and Fig. 9.5*b*; we have neglected the effect of the resistance R_g in comparison with the tuned-circuit conductance g_c. This circuit, in turn can be represented by the equivalent circuit in Fig. 9.5*c*, which finally corresponds to Fig. 9.5*d*. The Barkhausen condition is then

$$\frac{(g_m M/L)v_g}{1/(j\omega L) + j\omega C + g_c} = v_g \tag{9.11}$$

Multiplying both sides by $1/(j\omega L) + j\omega C + g_c$ and dividing by v_g yields

$$g_m \frac{M}{L} = \frac{1}{j\omega L} + j\omega C + g_c \tag{9.12}$$

Equating the real and imaginary parts of Eq. (9.12) gives

$$\frac{1}{j\omega L} + j\omega C = 0 \qquad \text{or} \qquad \omega^2 LC = 1 \qquad g_m \frac{M}{L} = g_c \tag{9.12a}$$

The first equation gives the *frequency of oscillation*. We see that the frequency changes only by a factor of 3 when C changes by a factor of 9, whereas in the Wien-bridge oscillator the frequency changed by a factor of 9 under these conditions. The second condition gives the *amplitude of oscillation*. In this circuit g_m decreases with increasing amplitude v_g. If $g_m M/L > g_c$ for small amplitudes, v_g increases with time until a stable amplitude is reached for which

$$g_m(v_g)\frac{M}{L} = g_c \tag{9.12b}$$

Equation (9.12) can also be written as

$$\frac{1}{j\omega L} + j\omega C + g_c - g_m \frac{M}{L} = 0 \tag{9.13}$$

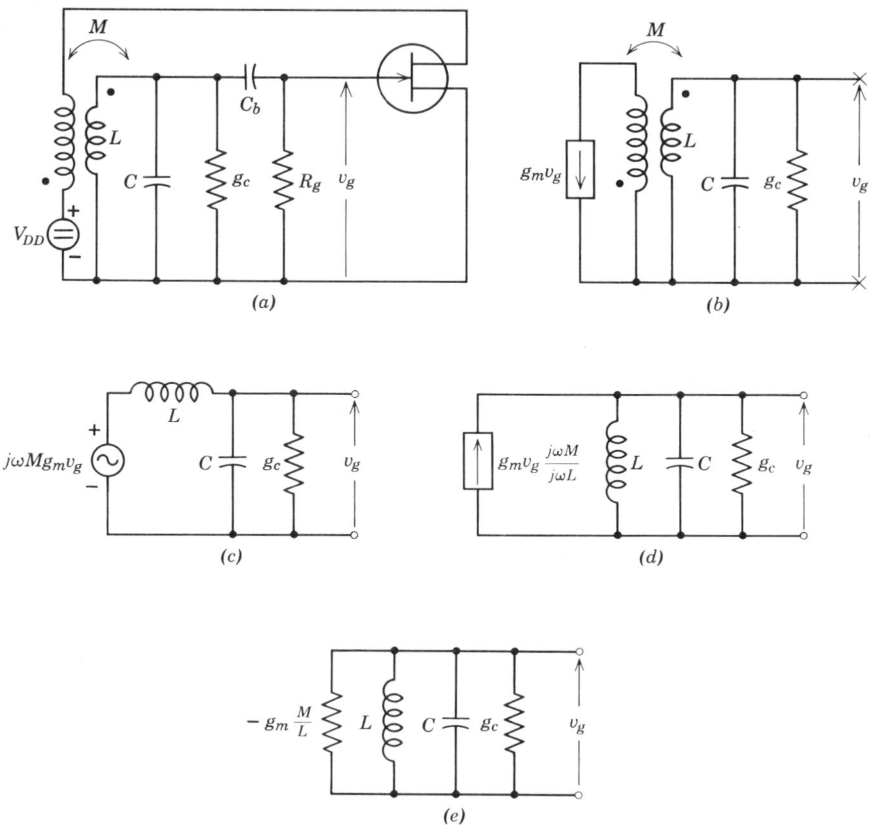

Figure 9.5. Magnetically coupled feedback oscillator. (*a*) The circuit. (*b*) Simplified equivalent circuit. (*c*) Further simplified equivalent circuit. (*d*) Still further equivalent circuit. (*e*) Equivalent circuit with a negative conductance.

This is easily understood with the help of Fig. 9.5*d*. The current generator $(g_m M/L)v_g$ in Fig. 9.5*d* that establishes v_g corresponds to a feedback conductance $-g_m M/L$ in parallel with the tuned circuit (see Fig. 9.5*e*). In Section 9.2b we derive Eq. (9.13) from another point of view.

Example 3. Find the value of the capacitance C for an oscillator based on Fig. 9.5*a* which generates 1 MHz for an inductance of 10^{-4} H.

Answer.

$$C = \frac{1}{\omega^2 L} = \frac{1}{4\pi^2 \times 10^{12} \times 10^{-4}} \cong \frac{10^{-9}}{4} = 250 \text{ pF}$$

Example 4. The FET used in the oscillator has a transconductance $g_m = 10^{-3}$ mhos for small signals, whereas the tuned-circuit conductance is 10^{-4} mhos. Find the value of M/L so that

$$g_m \frac{M}{L} = 4g_c$$

for small signals.

Answer.

$$\frac{M}{L} = \frac{4g_c}{g_m} = \frac{4 \times 10^{-4}}{10^{-3}} = 0.4$$

Example 5. In an oscillator with variable C the minimum circuit capacitance is 30 pF and the maximum circuit capacitance is 270 pF, so the frequency of oscillation can change by a factor of 3. Find the value of L for which the frequency of oscillation can change from 1 MHz to 3 MHz.

Answer. The lowest frequency is obtained for the largest value of C. Therefore

$$L = \frac{1}{\omega^2 C} = \frac{1}{4\pi^2 \times 10^{12} \times 270 \times 10^{-12}} = \frac{1}{4\pi^2 \times 270} = 0.94 \times 10^{-4} \text{ H}$$

9.2 ADMITTANCE AND IMPEDANCE CONSIDERATIONS IN OSCILLATORS

There are cases in which the feedback consideration discussed in Section 9.1 does not apply. Certain solid-state devices, such as tunnel diodes, have a negative conductance. If such a device is connected in parallel with a tuned circuit, oscillations can occur, even though no feedback exists. However, feedback circuits that produce oscillations can also be represented by an input admittance Y_{in} or an input impedance Z_{in}, each with a negative real part. It is therefore necessary to consider oscillators from an impedance or admittance point of view also.

9.2a Transient Response of Tuned Circuits

Figure 9.6*a* shows an L–C–g parallel circuit with g either positive or negative. We now investigate the transient response of this circuit by

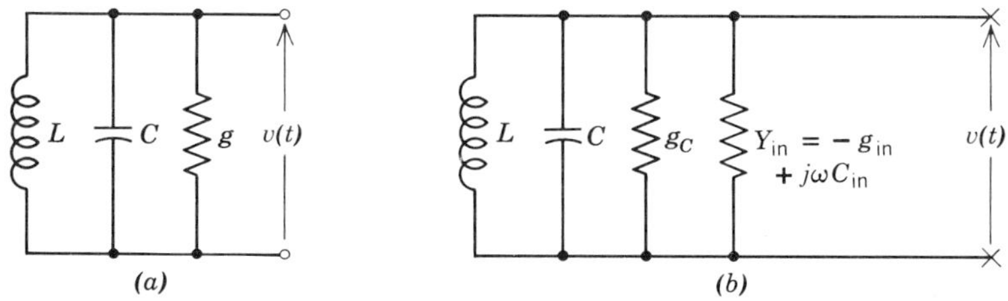

Figure 9.6. (*a*) *L*-*C*-*g* circuit. (*b*) *L*-*C*-g_c circuit with an admittance Y_{in}, having a negative conductance, in parallel.

finding the general solution of the homogeneous integrodifferential equation

$$C\frac{dv}{dt}+gv+\frac{1}{L}\int v\,dt=0 \tag{9.14}$$

This can be done by substituting $v=\exp(st)$ and solving for s. When this has been accomplished, the general solution can be written in the form

$$v=A\exp\left(-\frac{1}{2}\frac{g}{C}t\right)\cos\left\{\omega_0 t\left[1-\left(\frac{g}{2\omega_0 C}\right)^2\right]^{1/2}+\varphi\right\} \tag{9.14a}$$

for $|2\omega_0 C/g|>1$, whereas for $|2\omega_0 C/g|<1$,

$$v=A\exp\left(-\frac{1}{2}\frac{g}{C}t\right)\cosh\left\{t\left[\left(\frac{g}{2C}\right)^2-\omega_0^2\right]^{1/2}+\varphi\right\} \tag{9.14b}$$

Here the frequency ω_0 is defined by

$$\omega_0^2 LC=1 \tag{9.14c}$$

We thus have the following possibilities.

1. $g>2\omega_0 C$. The solution dies out aperiodically.
2. $0<g<2\omega_0 C$. The solution is a damped oscillation.
3. $g=0$. The solution is an undamped oscillation of frequency ω_0.
4. $-2\omega_0 C<g<0$. The solution is an oscillation with an exponentially growing amplitude.
5. $g<-2\omega_0 C$. The solution is nonoscillatory and grows exponentially.

Since for an undamped oscillation $g=0$ and $\omega_0^2 LC=1$, we can write, for $\omega=\omega_0$,

$$Y_{\text{par}}=g+\frac{1}{j\omega_0 L}+j\omega_0 C=0 \tag{9.15}$$

Figure 9.6*b* shows an actual L–g_c–C parallel circuit connected in parallel with an admittance, $Y_{\text{in}}=-g_{\text{in}}+j\omega C_{\text{in}}$, with a negative real part, caused by feedback. Applying the previous discussion, we find for the condition for oscillation:

$$Y_{\text{par}}=g_c+\frac{1}{j\omega_0 L}+j\omega_0 C-g_{\text{in}}+j\omega_0 C_{\text{in}}=0 \tag{9.16}$$

Equation (9.16) implies that both the real and imaginary parts of Y_{par} must be zero. This yields

$$\frac{1}{j\omega_0 L}+j\omega_0(C+C_{\text{in}})=0 \quad \text{or} \quad \omega_0=\left[L(C+C_{\text{in}})\right]^{-1/2} \qquad g-g_{\text{in}}=0 \tag{9.16a}$$

The first condition gives the frequency of oscillation, and the second condition gives the amplitude v of the oscillations. To start the oscillations, $g_c-g_{\text{in}}(v)$ must be negative for small values of v. The oscillations then start and grow in amplitude. Because of the nonlinearity of the device, $g_{\text{in}}(v)$ ultimately decreases with increasing v. The amplitude v of the oscillations thus grows until $g_c-g_{\text{in}}(v)=0$. When this happens, a stable amplitude has been attained.

We now apply similar conditions to an L–R–C series circuit (Fig. 9.7*a*), where R may again be positive or negative, and investigate the transient response of the network by solving the homogeneous integrodifferential

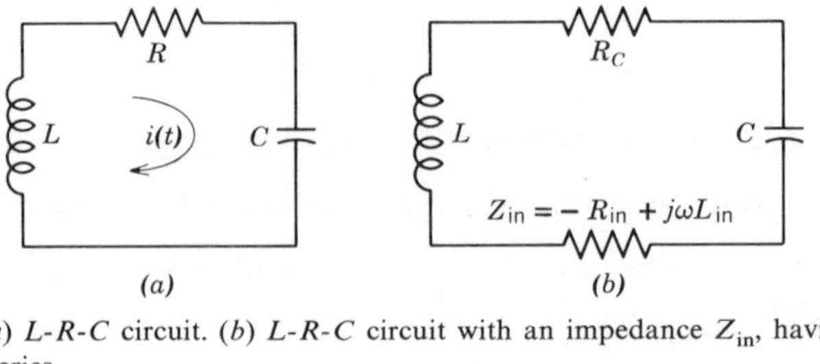

Figure 9.7. (*a*) *L-R-C* circuit. (*b*) *L-R-C* circuit with an impedance Z_{in}, having a negative resistance, in series.

equation

$$L\frac{di}{dt} + Ri + \frac{1}{C}\int i\,dt = 0 \tag{9.17}$$

Considering the same possibilities as before, we find that a stable oscillation of frequency $\omega_0 = (LC)^{-1/2}$ is obtained if $R = 0$. This corresponds to the equation

$$Z_{\text{loop}} = j\omega_0 L + \frac{1}{j\omega_0 C} + R = 0 \tag{9.18}$$

Figure 9.7*b* shows an actual L–R_c–C series circuit applied to an impedance, $Z_{\text{in}} = -R_{\text{in}} + j\omega L_{\text{in}}$ with a negative real part, caused by feedback. Equation (9.18) then yields

$$Z_{\text{loop}} = j\omega_0 L + \frac{1}{j\omega_0 C} + R_c - R_{\text{in}} + j\omega_0 L_{\text{in}} = 0 \tag{9.19}$$

Setting the real part equal to zero and the imaginary part equal to zero yields

$$j\omega_0 L + \frac{1}{j\omega_0 C} + j\omega_0 L_{\text{in}} = 0 \quad \text{or} \quad \omega_0 = \left[(L + L_{\text{in}})C\right]^{-1/2} \qquad R_c - R_{\text{in}}(i) = 0 \tag{9.20}$$

The first condition again gives the frequency of the oscillations, and the second condition gives the (current) amplitude. To start the oscillations, one must require that $R_c - R_{\text{in}}(i) < 0$ for small amplitudes. The amplitude i of the oscillations then grows; this ultimately causes a decrease in $R_n(i)$, so that finally a stable amplitude is reached for which $R_c - R_{\text{in}}(i) = 0$.

9.2b Applications

As a first example we discuss the circuit in Fig. 9.5*a* from an admittance point of view. Since a parallel tuned circuit is involved, the condition for oscillation is $Y_{\text{par}} = 0$. To apply this condition, the admittance Y_{in} caused by the feedback must be evaluated. To this end, C and g_c in Fig. 9.5*a* are replaced by an emf v_g, and the FET is replaced by a current generator $g_m v_g$; this gives the circuit in Fig. 9.8*a*. It is easily seen by inspection that the current generator $g_m v_g$ gives rise to a current $i = -g_m v_g j\omega M / j\omega L$. Therefore Y_{in} is

$$Y_{\text{in}} = \frac{i}{v_g} = -g_m\frac{M}{L} \tag{9.21}$$

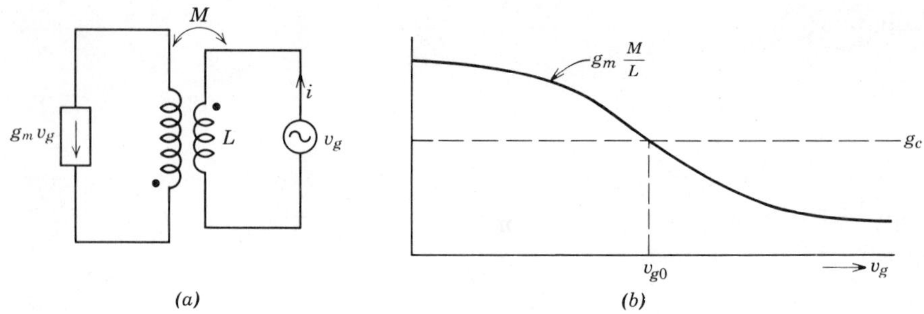

Figure 9.8. (*a*) Part of the circuit equivalent to Fig. 9.5 redrawn, showing how the negative input conductance is calculated. (*b*) $g_m M/L$ and g_c plotted versus the gate amplitude v_g. The point where these curves intersect gives the stable amplitude of oscillation.

and from Eq. (9.16) the condition for oscillation can be written as

$$g_c + j\omega_0 C + \frac{1}{j\omega_0 L} - g_m \frac{M}{L} = 0 \tag{9.22}$$

which corresponds exactly to Eq. (9.13).

To understand the stable operating point, we have to determine how the transconductance g_m depends on the gate voltage v_g. To this end we return to Fig. 9.5*a* and observe that the gate of the FET begins to rectify if the amplitude of the oscillations increases. This effectively clamps the maximum gate voltage $V_{g,\max}$ to the turn-on voltage of the gate diode, which is about 0.70 V. Because of the gate current that flows, the dc bias of the gate shifts further negative until the FET passes current only part of the time. The current $I(t)$ flowing through the FET then has a fixed peak value, and the width of the current pulse decreases with increasing amplitude. If we now write

$$I(t) = I_0 + i_1 \cos\omega_0 t + i_2 \cos 2\omega_0 t + \cdots \tag{9.23}$$

we have that

$$i_1 = \frac{1}{\pi}\int_{-\pi}^{\pi} I(t)\cos\omega_0 t\, d(\omega_0 t) \tag{9.23a}$$

decreases with increasing amplitude v_g. Since the transconductance g_m must be defined as

$$g_m = \frac{i_1}{v_g} \tag{9.24}$$

we see that g_m decreases with increasing amplitude v_g.

If $g_m M/L$ is plotted versus v_g, we thus obtain the curve shown in Fig. 9.8*b*. Plotting the curve for g_c also, we see that a stable amplitude v_{g0} is obtained when

$$g_m(v_{g0})\frac{M}{L}=g_c \tag{9.24a}$$

This amplitude is stable, since if for some reason v_g increases, $-g_m M/L+g_c$ will become positive, and this would tend to decrease the amplitude. Or if v_g decreases, $-g_m M/L+g_c$ will become negative, and this would tend to increase the amplitude.

As a second example we consider the circuit in Fig. 9.9*a*, where jX_1 and jX_2 are reactances of the same sign. If both are capacitive reactances, jX_2 must be shunted by a choke or resistor to allow passage of emitter current. If they are inductive reactances, a blocking capacitor must be put in series with jX_1, otherwise the base will be short-circuited to the emitter.

Since this is a series tuned circuit, the condition for oscillation is $Z_{\text{loop}}=0$. To evaluate the input impedance Z_{in} provided by the feedback circuit, we replace the series tuned circuit by an emf v, replace jX_1 by $Y_1=1/jX_1$ and jX_2 by $Y_2=1/jX_2$, and obtain the circuit in Fig. 9.9*b*. If $X_1 \ll r_\pi$, we obtain for the current i

$$i=(v-v_e)Y_1 \qquad v_e Y_2=g_m(v-v_e)+Y_1(v-v_e) \tag{9.25}$$

Solving for v_e yields

$$v_e=\frac{g_m+Y_1}{g_m+Y_1+Y_2}v \qquad v-v_e=\frac{Y_2}{g_m+Y_1+Y_2}v \tag{9.26}$$

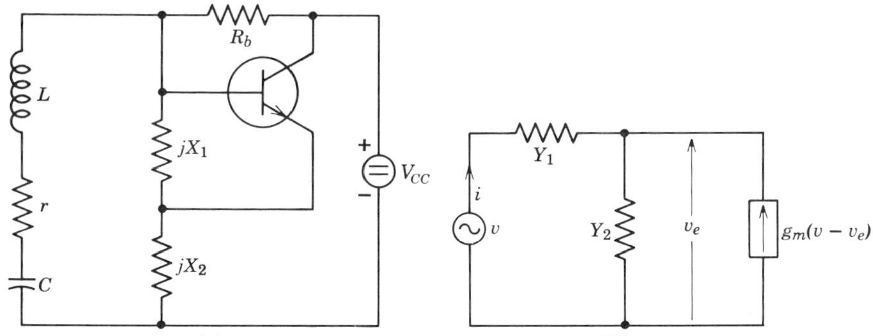

Figure 9.9. Transistor oscillator with a very stable frequency. (*a*) The actual circuit. (*b*) The equivalent circuit.

so

$$i = \frac{Y_1 Y_2}{g_m + Y_1 + Y_2} v \tag{9.27}$$

Consequently,

$$Z_{in} = \frac{v}{i} = \frac{g_m + Y_1 + Y_2}{Y_1 Y_2} = \frac{g_m}{Y_1 Y_2} + \frac{1}{Y_1} + \frac{1}{Y_2} = j(X_1 + X_2) - g_m X_1 X_2 \tag{9.28}$$

From Eq. (9.19) the condition for oscillation is therefore

$$\frac{1}{j\omega_0 C} + r + j\omega_0 L + j(X_1 + X_2) - g_m X_1 X_2 = 0 \tag{9.29}$$

so

$$\frac{1}{j\omega_0 C} + j\omega_0 L + j(X_1 + X_2) = 0 \qquad r - g_m X_1 X_2 = 0 \tag{9.29a}$$

The first equation determines the frequency of oscillation, and the second determines the amplitude, because g_m again decreases with increasing amplitude. The merit of the circuit is that X_1 and X_2 can be quite small, because g_m is large, so $j(X_1 + X_2)$ is small. Therefore the frequency of the oscillation is hardly affected by the properties of the transistor. Consequently, if the series tuned circuit is kept at a fixed temperature, an extremely stable frequency is generated. At lower frequencies X_1 and X_2 are usually chosen inductive, and at higher frequencies they are capacitive.

Example 1. An oscillator at 1 MHz has $C = 100$ pF and $g_m = 10^{-1}$ mho, and the tuned circuit has a Q factor $Q_0 = 100$. If $X_1 = X_2 = X$, find the value of X needed for oscillations to occur.

Answer.

$$\frac{1}{Q_0} = \omega_0 C r \qquad r = \frac{1}{\omega_0 C Q_0} = \frac{1}{2\pi \times 10^6 \times 10^{-10} \times 100} = 16\ \Omega$$

$$|X_1| = |X_2| = |X| = \left(\frac{r}{g_m}\right)^{1/2} = \left(\frac{16}{0.1}\right)^{1/2} = 12.6\ \Omega$$

so the condition $|X_1| \ll r_\pi$ is well satisfied. If the reactances are capacitive, we have

$$X_1 = X_2 = -\frac{1}{\omega_0 C_0} \qquad C_0 = \frac{1}{\omega_0 |X|} = \frac{1}{2\pi \times 10^6 \times 12.6} = 12{,}600\ \text{pF}$$

9.3 LASERS

A laser is an oscillator oscillating at optical frequencies. We show here that the theory discussed in Section 9.2, if properly modified, can be applied to this case. It is beyond the scope of this book to discuss individual lasers in detail.

Let us consider an optical cavity consisting of two flat mirrors at a distance L, with light reflecting back and forth between them parallel to the axis of the cavity (Fig. 9.10*a*). Optical resonance occurs if

$$2L = n\lambda = \frac{nv}{f_0} \qquad \text{or} \qquad f_0 = \frac{nv}{2L} \tag{9.30}$$

where n is integral ($n = n_0, n_0 + 1, \ldots$) and v is the phase velocity of the waves. This cavity corresponds to a tuned circuit.

Now let material be located between the two mirrors, with atoms having energy levels E_1 and E_2 ($E_2 > E_1$) and let N_1 atoms per cubic centimeter of energy E_1 and N_2 atoms per cubic centimeter of energy E_2 be present in the material. Then light of frequency f (Fig. 9.10*b*)

$$f = \frac{E_2 - E_1}{h} \tag{9.31}$$

will be absorbed or emitted. This equation implies that light is absorbed or emitted as quanta hf, where h is Planck's constant.

At equilibrium $N_2 < N_1$ but, if equilibrium is sufficiently disturbed, we may have $N_2 > N_1$. This is called *population inversion*.

The emitted line has a certain frequency range Δf, so optical resonance may occur for several frequencies f_0 within the emission line. The frequency separation of these resonances is therefore

$$\Delta f_0 = \frac{v}{2L} \tag{9.32}$$

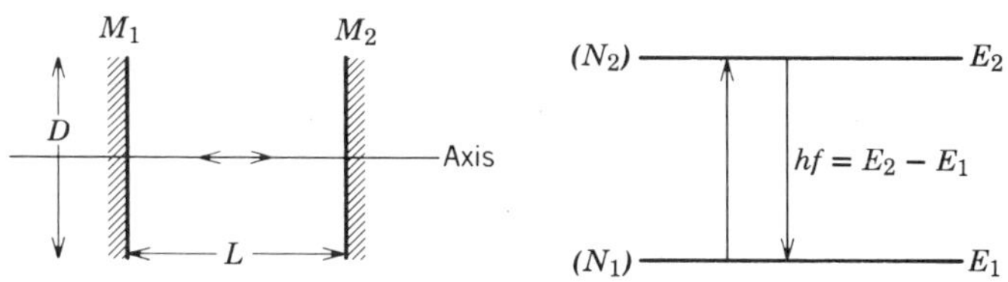

Figure 9.10. (*a*) Optical cavity consisting of two mirrors at a distance L. (*b*) Energy-level diagram of the active material in the cavity.

These frequencies are called *modes of oscillation*.

Let an energy density $u(f)$ in a particular mode of oscillation be present in the cavity; then there are two possible processes:

1. *Absorption*, by which atoms in the state E_1, under the influence of the radiation field, are brought to the state E_2 through absorption of quanta hf. The rate of transitions per second is proportional to both $u(f)$ and N_1, so it can be written as

$$Cu(f)N_1 \tag{9.33}$$

where C is a proportionality factor.

2. *Stimulated emission*, by which atoms in the state E_2, under the influence of the radiation field, are brought to the state E_1 through emission of quanta hf. The rate of transitions per second is

$$Cu(f)N_2 \tag{9.34}$$

where C is the same factor as in Eq. (9.33).

The net effect is thus a rate of increase in $u(f)$ with time

$$\frac{du(f)}{dt} = hfCu(f)(N_2 - N_1) \tag{9.35}$$

since each transition adds or subtracts a quantum hf to the stored energy. For $N_2 < N_1$ the energy density $u(f)$ present in the cavity decays by absorption, and for $N_2 > N_1$ (population inversion) the energy density $u(f)$ rises exponentially until the condition $N_2 = N_1$ is reached. The condition for oscillation for this lossless case is therefore

$$N_2 \geqslant N_1 \tag{9.35a}$$

Lasers differ in the way in which the population inversion is maintained.

Now we take optical losses into account. Let the medium present in the optical cavity have a power loss coefficient α_m per centimeter for traveling waves. Furthermore, let one of the two mirrors have a reflection coefficient R, or a transmission coefficient $1 - R$, so that the mirror transmits part of the generated radiation; as a consequence the effective power loss coefficient of the system is $(1 - R)/2L$ per centimeter. If we multiply by the group velocity v_g of the waves, the rate of energy loss per unit time is $u(f)v_g[\alpha_m + (1 - R)/2L]$. We thus have, instead of Eq. (9.35),

$$\frac{du(f)}{dt} = u(f)\left[hfC(N_2 - N_1) - \left(\alpha_m + \frac{1-R}{2L}\right)v_g\right] \tag{9.36}$$

Oscillations will now start if

$$hfC(N_2 - N_1) - \left(\alpha_m + \frac{1-R}{2L}\right)v_g > 0 \tag{9.36a}$$

N_2 decreases with increasing $u(f)$, and N_1 increases with increasing $u(f)$, so a stable energy density u_0 is maintained if

$$hfC[N_2(u_0) - N_1(u_0)] - \left(\alpha_m + \frac{1-R}{2L}\right)v_g = 0 \tag{9.36b}$$

This means that a stable amount of optical power is transmitted through the partly transparent mirror.

The light of a laser is emitted in a very narrow beam. The reason is that light propagating in a direction accurately parallel to the axis of the optical system is regeneratively amplified only if it has the same phase at successive bounces, that is, if Eq. (9.30) is satisfied. All other light waves are suppressed by destructive interference or run out of the cavity.

The directionality of the emitted beam is limited by diffraction of the beam at the partly transparent mirror. The theory of diffraction shows that the angular divergence $\delta\theta$ of the energy beam is

$$\delta\theta = \frac{v}{f_0 D} \tag{9.37}$$

where D is the diameter of the mirror, f_0 the frequency, and v is the phase velocity of the waves.

For long cavities Eq. (9.30) is satisfied for more than one frequency in the emitted line. To avoid the generation of multiple frequencies one should either make the cavity short [see Eq. (9.32)] or suppress the undesired frequencies optically.

PROBLEMS

9.1 Figure 9.11 shows a tuned drain oscillator. Apply the Barkhausen condition and show that the condition for oscillation is

$$Y_{\text{par}} = \frac{1}{R} + \frac{1}{j\omega_0 L} + j\omega_0 C - g_m \frac{M}{L} = 0$$

Ignore the effects of R_g and C_b.

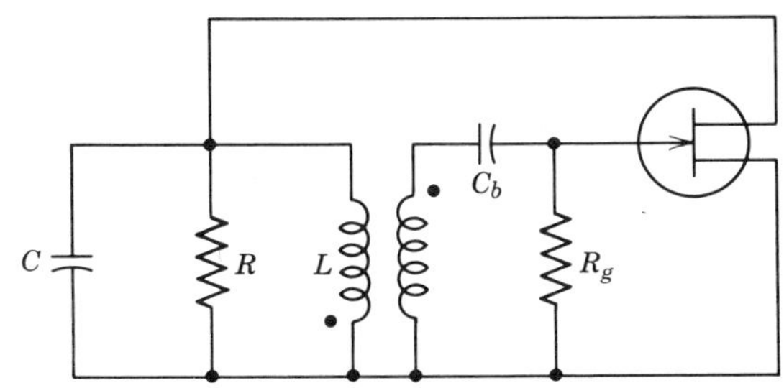

Figure 9.11. Tuned drain oscillator.

9.2 Figure 9.12 shows the Colpitts oscillator. Apply the Barkhausen condition and show that the condition for oscillation yields

$$\omega_0^2 L \frac{C_1 C_2}{C_1 + C_2} = 1 \qquad g_m = \omega_0^2 C_1 C_2 r$$

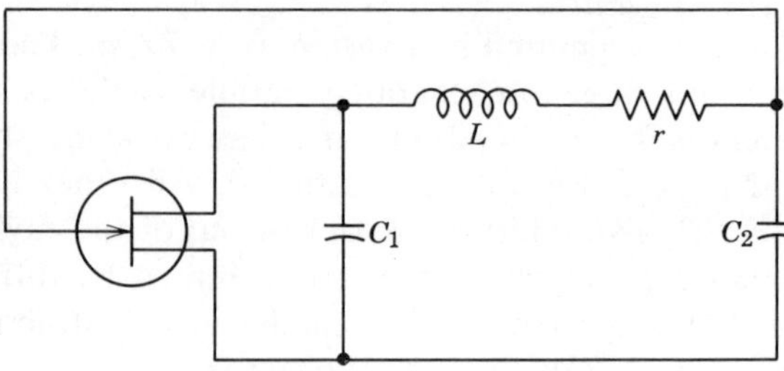

Figure 9.12. Colpitts oscillator.

9.3 Figure 9.13 shows the Hartley oscillator. Find the condition for oscillation by applying the Barkhausen condition and show that

$$\omega_0^2 (L_1 + L_2) C = 1 \qquad g_m = \frac{r_1 + r_2}{\omega_0^2 L_1 L_2}$$

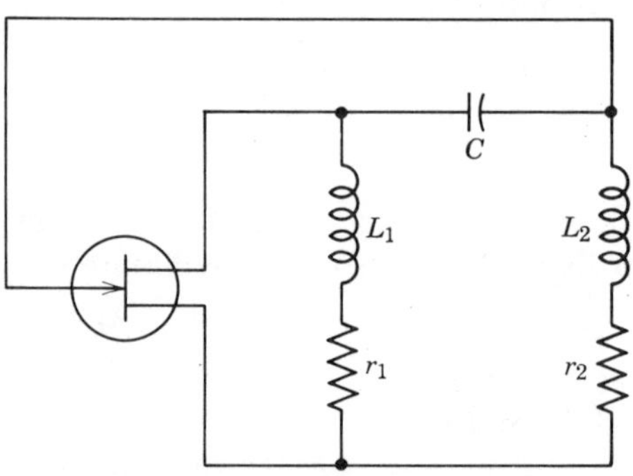

Figure 9.13. Hartley oscillator.

9.4 Figure 9.14 has $\omega^2 C_{dg} L_g \ll 1$ for all frequencies of practical interest. If the device has a transconductance g_m, show that the condition for oscillation is

$$Y_{\text{par}} = \frac{1}{j\omega_0 L} + \frac{1}{R} + j\omega_0 (C + C_{dg}) - g_m \omega_0^2 L_g C_{dg} = 0$$

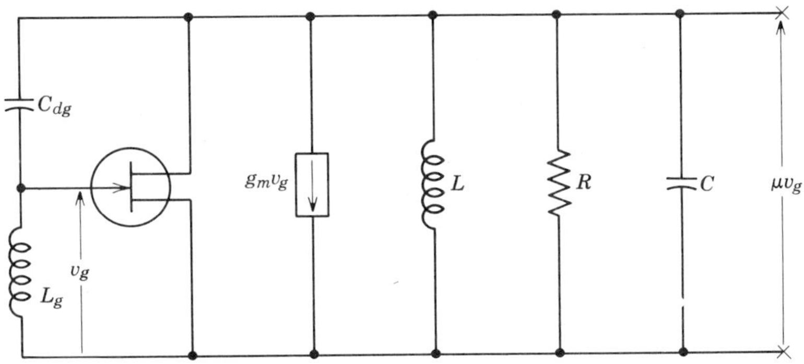

Figure 9.14. *L-C* phase-shift oscillator.

9.5 Prove the result from Problem 9.4 with the help of the Barkhausen condition.

9.6 The phase-shift oscillator in Fig. 9.15 has a three-stage *CR* phase-shift network. If the amplifier has the output voltage v_0 180° out of phase with respect to the input voltage v_i, the output impedance of the amplifier is very low, and the input impedance very high, show that the condition for oscillation is

$$|g_v| = 29 \qquad \omega_0 CR = \frac{1}{6^{1/2}}$$

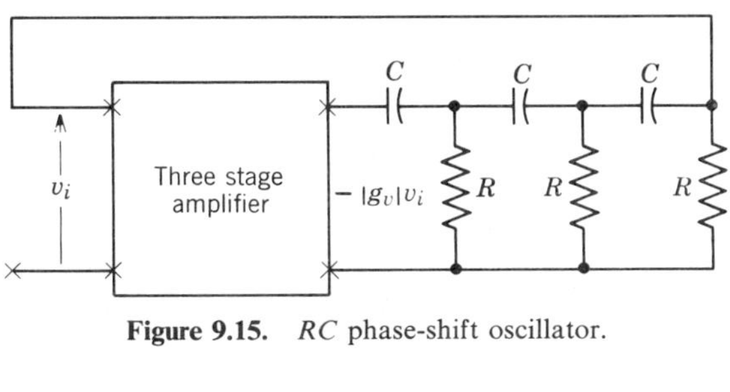

Figure 9.15. *RC* phase-shift oscillator.

9.7 A tunnel diode oscillator has a circuit as shown in Fig. 9.16. At the operating point the device gives a negative conductance

$$-\frac{g_{d0}}{1+(v/v_1)^2}$$

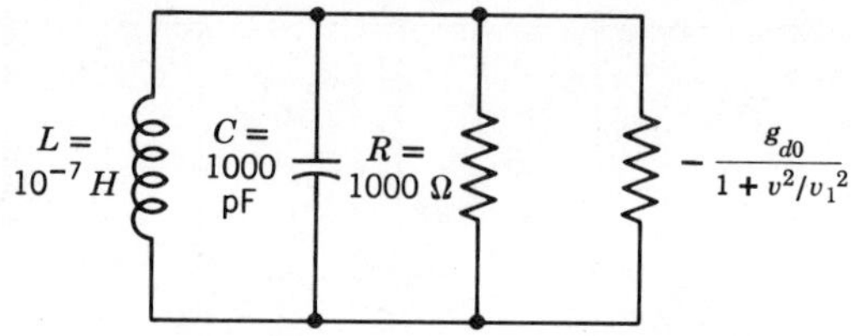

Figure 9.16. Tunnel diode oscillator.

where v is the amplitude of the oscillation. Here $g_{d0}=10^{-2}$ Ω and $v_1=$ 0.05 V.

(a) Write down the condition for oscillation.
(b) Find the frequency of oscillation.
(c) Find the stable amplitude of oscillation.

9.8 The circuit in Fig. 9.17 shows a bridge-type oscillator which performs better than the corresponding bridge circuit discussed in the text. If $g_v=30$, find the value of R_2 for which the circuit just oscillates at the frequency $\omega_0=1/CR$. C_b is a blocking capacitor of negligible impedance.

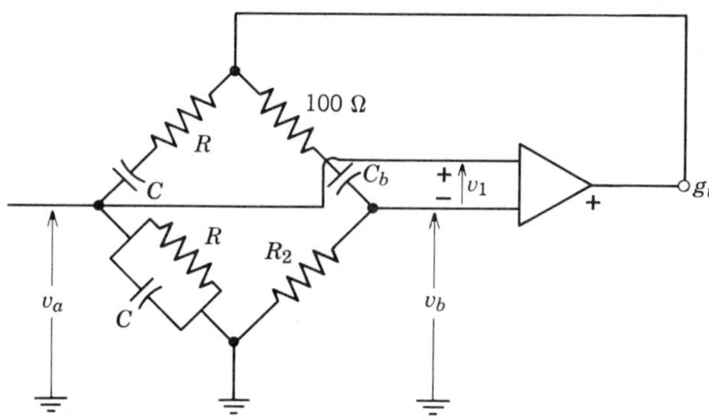

Figure 9.17. Bridge-type oscillator.

9.9 A two-stage FET oscillator has the phase-shift network shown in Fig. 9.18. Find the condition of oscillation, involving the minimum value of μ and the frequency.

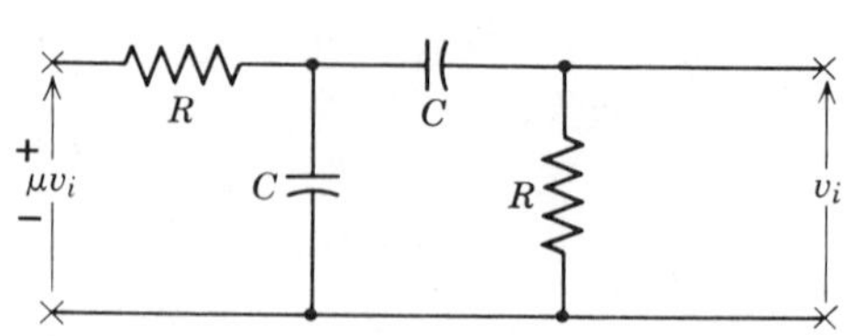

Figure 9.18. *RC* phase-shift oscillator circuit.

9.10 A two-stage FET oscillator has the phase-shift network shown in Fig. 9.19. Find the condition of oscillation, involving the minimum value of μ and the frequency.

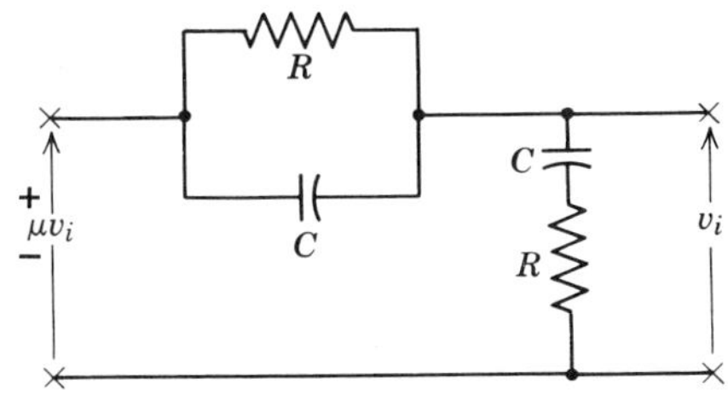

Figure 9.19. Another RC phase-shift oscillator.

CHAPTER

10 Storage and Transmission of Information

This chapter introduces various circuits used for storage and transmission of information, all of which use some type of flip-flop circuit.

Section 10.1 discusses transistorized flip-flops and indicates how clock pulses can be applied to the circuit. Section 10.2 deals with various memory circuits and latches, such as nand and nor gate latches, gated memory and data latches (D-type flip-flops), T-type flip-flops, R-S flip-flops, J-K flip-flops, two types of master-slave flip-flops, and a MOSFET realization of one of them. Section 10.3 briefly considers shift registers and counters.

10.1 THE TRANSISTORIZED FLIP-FLOP

We return to the transistor bistable circuit discussed in Chapter 5 but, instead of connecting the trigger pulse directly, we feed the trigger pulses through a small capacitor. The CR input circuit then differentiates the pulse, giving rise to a positive pulse of height V at the leading edge of the pulse transmitted through the diode that triggers the circuit and to a negative pulse of height V at the trailing edge of the pulse not transmitted by the diode (Fig. 10.1).

If Q is at a logical 0 and a trigger pulse is applied to S, this turns Q_2 on and Q_1 off, so Q switches to a logical 1. If Q is at a logical 1 and a trigger pulse is applied to S, nothing happens, since Q_2 stays on. If Q is at a

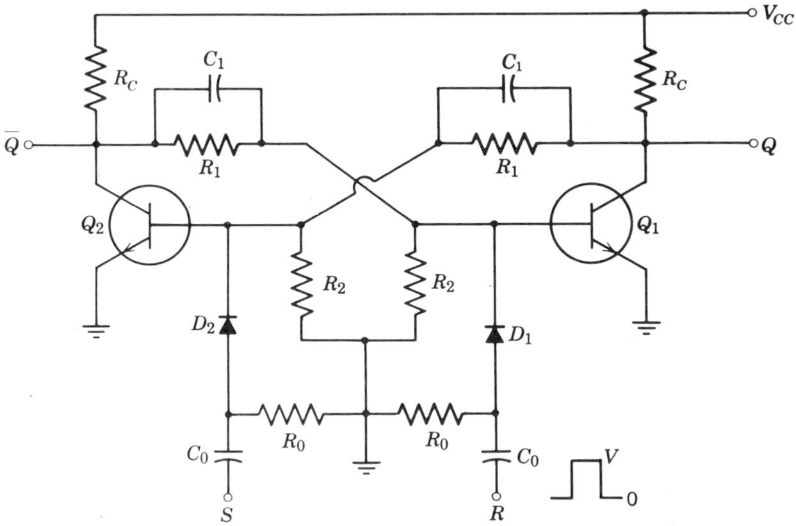

Figure 10.1. Transistorized flip-flop with differentiating circuit at input.

logical 0 and a trigger pulse is applied to R, nothing happens, for Q_1 stays on. If Q is at a logical 1 and a trigger pulse is applied to R, Q_1 is switched on, hence Q switches to a logical 0. The pulse at S thus sets Q at 1 or keeps it in that state; the pulse at R sets Q at 0 or keeps it in that state. For that reason the two inputs are also called *set* and *clear* or *reset*, respectively.

Next we modify the circuit by adding an opportunity to apply clock pulses to the circuit. The circuit thus obtained is called an *RS* flip-flop (Fig. 10.2).

We now explain the operation of this circuit. If a clock pulse is applied, and S is at 0 V (logical 0), the leading edge of the clock pulse produces by differentiation a positive voltage $+V$ across the 10-kΩ resistor, which is clipped by diode D_4 and not transmitted through D_2. The trailing edge of the clock pulse produces by differentiation a negative voltage $-V$ across the 10-kΩ resistor, which is not clipped by diode D_4 but is transmitted through D_2. If Q_2 was first on, it turns Q_2 off and, if Q_2 was already off, it stays off.

If a clock pulse is applied and S is at the voltage V (logical 1), nothing happens, since the voltage across the 10-kΩ resistor can never become negative, so Q_2 stays in the state it was. Similar considerations hold for the voltage at R.

We thus have the following truth table. t_n refers to the input state *before* the clock pulse is applied, and t_{n+1} refers to the output state after the clock pulse has been applied. The second, third, and fourth statements of the

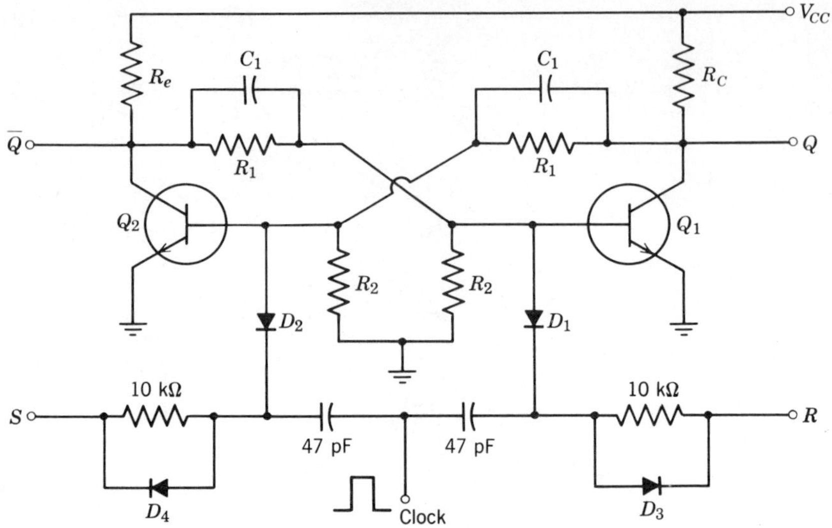

Figure 10.2. Transistorized flip-flop with provision for applying clock pulses through a differentiating circuit. (From H. Malmstadt and C. G. Enke, *Digital Electronics for Scientists,* Copyright © 1969, W. A. Benjamin, Inc., Menlo Park, Cal.)

truth table speak for themselves. If $S=0$ and $R=1$, a negative pulse is transmitted through D_2, which turns Q_2 off when on and leaves it off when off, so Q is a 0 after clocking. If $S=1$ and $R=0$, a negative pulse is transmitted through D_1, which turns Q_1 off when on and leaves it off when off, so Q is a 1 after clocking. If $R=1$ and $S=1$, no negative pulse is transmitted through either diode D_1 or D_2, so the state of Q is not changed. If $R=0$ and $S=0$, a negative pulse is transmitted through *both* diodes and the situation is indeterminate.

Input, t_n		Output, t_{n+1}
S	R	Q
0	0	Indeterminate
0	1	0
1	0	1
1	1	$Q_n(=Q \text{ at } t_n)$

We then have that, for $R \neq S$, Q is set to a 0 or a 1, independent of its previous state, whereas, for $R=S=1$, the output stays as it is, that is, it remembers its previous state.

10.2 MEMORY CIRCUITS AND LATCHES

10.2a Nand and Nor Gate Latches

Figure 10.3 shows the nand latch. To understand the circuit we observe that the gates G_1 and G_2 are nand gates, that is, they have a 1 output unless both inputs are 1s. Q and $\bar{Q}$ must be opposites, that is, if $Q=1$, then $\bar{Q}=0$, or if $Q=0$, then $\bar{Q}=1$. The state $Q=\bar{Q}$ is not allowed. We thus have the truth table

R	S		Q	$\bar{Q}$
0[a]	0[a]		1[a]	1[a]
0	1		1	0
1	0		0	1
1	1	Either	1	0
		or	0	1

[a] Not allowed.

If $R=0$, then $Q=1$, independent of the values of $\bar{Q}$ and S. This explains the second line of the truth table. For $S=0$, $\bar{Q}=1$, independent of the values of Q and R. This explains the third line of the truth table. For $R=0$ and $S=0$, we would have $\bar{Q}=Q=1$, but this state is not allowed.

The case $R=S=1$ requires special consideration. Suppose R is first at a 0 and S at a 1, and then R is switched to a 1. We had $Q=1$ and $\bar{Q}=0$ before switching, and nothing changes if R is switched to a 1, since the other input of gate G_1 is a 0. Suppose at first $R=1$ and $S=0$, and then S is switched to a 1. We had $Q=0$ and $\bar{Q}=1$ before switching, and nothing changes if S is switched to a 1 since the other input of gate G_2 is a zero. That is, if R was previously a 0, then $Q=1$ and $Q=0$; if S was previously

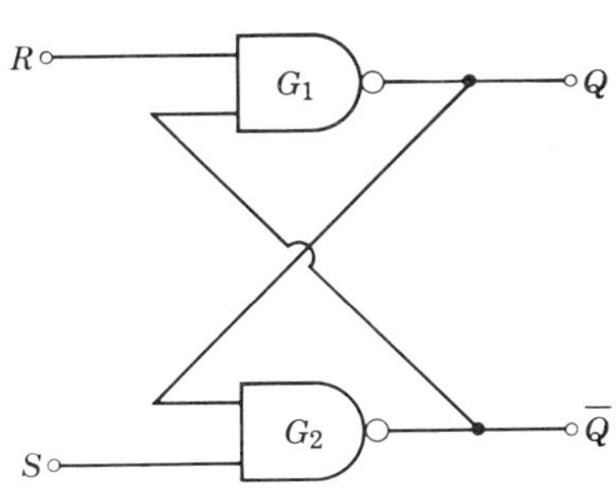

Figure 10.3. Nand latch.

a 0, then $Q=0$ and $\bar{Q}=1$. The circuit is therefore a *memory circuit*, it remembers the last 0 at either input.

The two cases $R \neq S$ gave a *fixed* response. Any R-S combination with $R \neq S$ sets the output in a prescribed manner. This is called *latching*, and the circuit is called a *nand latch*.

Figure 10.4 shows the *nor* latch. To explain the circuit, we observe that a nor circuit has a 1 output only when both inputs are 0. Q and $\bar{Q}$ must be opposite, that is, if $Q=1$, then $\bar{Q}=0$, and vice versa. The state $\bar{Q}=Q$ is not allowed. We thus have the truth table

Input			Output	
R	S		Q	$\bar{Q}$
0	0	Either	1	0
		or	0	1
1	0		0	1
0	1		1	0
1[a]	1[a]		0[a]	0[a]

[a] Not allowed.

If R is a 1, then Q is a 0, independent of the state of S and $\bar{Q}$. If $S=1$, $\bar{Q}=0$; for $S=0$, $\bar{Q}=1$. This explains the second line of the truth table. If S is a 1, then $\bar{Q}$ is a 0, independent of the state of R and Q. For $R=1$, $Q=0$; for $R=0$, $Q=1$. This explains the third line of the truth table. The case $R=S=1$ is not allowed. The case $R=S=0$ requires special consideration. Suppose R is first a 1 and S a 0, and then R is switched to a 0. We had $Q=0$ and $\bar{Q}=1$ before switching; nothing changes if R is switched, since the other input of gate G_1 is a 1. Suppose at first $R=0$ and $S=1$, and S is switched to a 0. We had $Q=1$ and $\bar{Q}=0$ before switching; nothing changes if S is switched, since the other input of gate G_2 is a 1. That is, if R was previously a 1, then $Q=0$ and $\bar{Q}=1$; if S was previously a 1, then $Q=1$

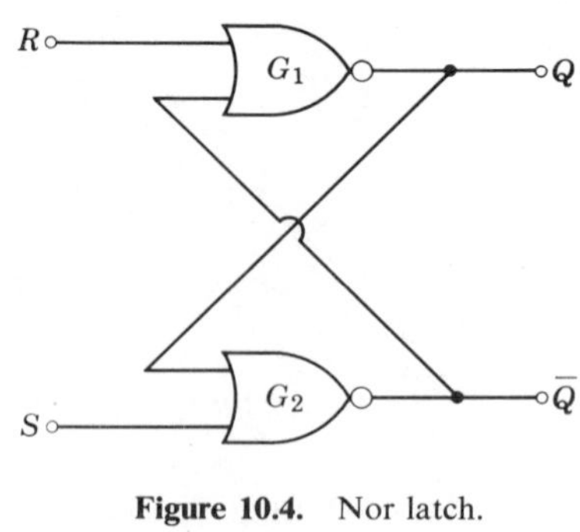

Figure 10.4. Nor latch.

and $\bar{Q}=0$. The circuit is thus a *memory* circuit; it remembers the last 1 at either input.

The two cases $R \neq S$ give a *fixed* response. Any R-S combination with $R \neq S$ sets the output in a prescribed manner. The circuit is therefore called a *nor latch*.

10.2b Gated Memory and Data Latches or *D*-Type Flip-Flops

Figure 10.5 shows a *gated memory circuit* with nand gates. It has the following truth table

Input, t_n (before clock pulse)		Output, t_{n+1} (after clock pulse)
S	R	Q
0	0	Q_n
0	1	0
1	0	1
1	1	Indeterminate

This follows from considering the operation of the input gate and the nand latch. Suppose $R=S=0$ and the clock is at 0 (before switching). Then $A=B=1$, hence $Q=0$ or 1, remembering the previous 0 at the input of the latch. If a clock pulse is now applied, we still have $A=B=1$, hence nothing changes at the output. The circuit then *remembers* its previous output. If $R \neq S$, the clock pulse *sets* Q in a prescribed manner. If $S=1$ and $R=0$ before clocking, $A=B=1$. When the clock pulse is applied, A switches to a 0, whereas B remains a 1, hence Q becomes a 1. If $S=0$ and $R=1$ before clocking, $A=B=1$. When the clock pulse is applied, B switches to a 0, whereas A remains a 1, hence Q becomes a 0, according to the truth table for Fig. 10.3.

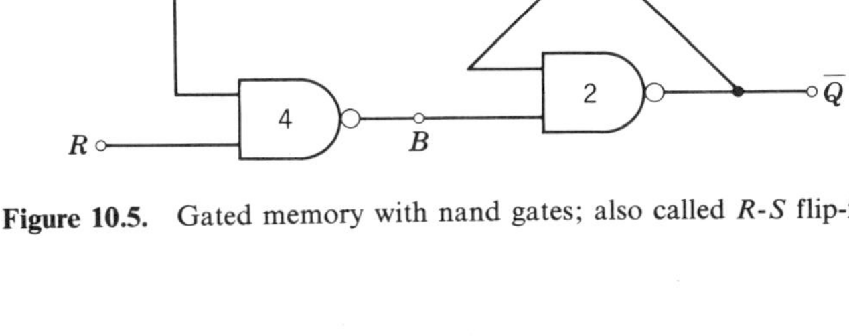

Figure 10.5. Gated memory with nand gates; also called R-S flip-flop.

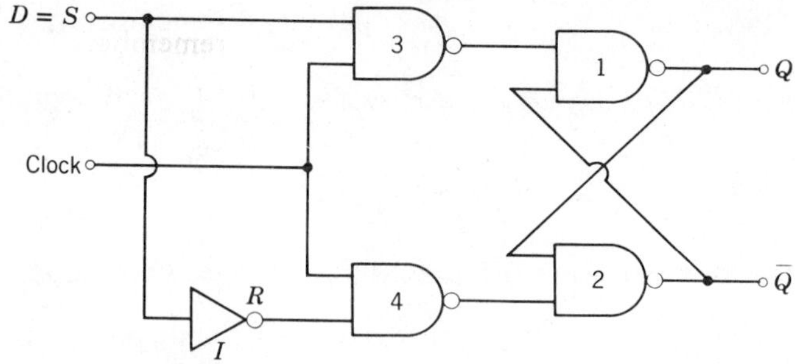

Figure 10.6. Data latch or D-type flip-flop.

A similar circuit, called a *data latch*, is shown in Fig. 10.6. The inverter I ensures that $R \neq S$. If $S=1$ and $R=0$, then according to the previous truth table $Q=1$, whereas for $S=0$ and $R=1$ we have $Q=0$. We then have the truth table

$D=\text{input}=t_n$	$Q=\text{output}=t_{n+1}$
0	0
1	1

In other words, the clock pulse transmits the input information to the output.

Instead of the term "data latch," the term "D-type flip-flop" is also used. One can also use a more complicated circuit, with additional set and clear inputs. They are overriding controls, that is, they determine the output state when activated. The symbol for such a configuration is shown in Fig. 10.7. The truth table is the same as for the previous case.

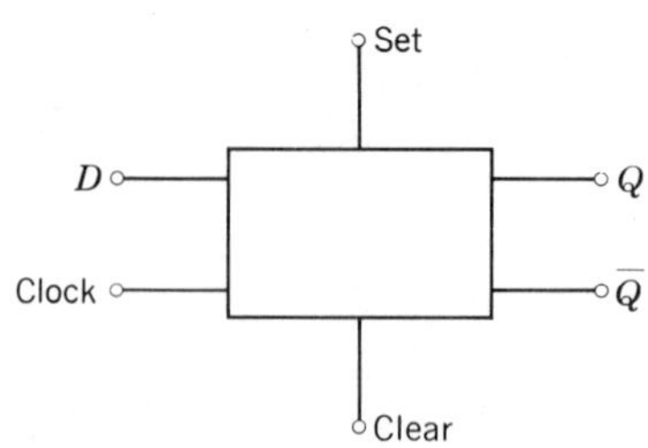

Figure 10.7. Symbol for data latch or D-type flip-flop.

10.2c The T-Type Flip-Flop

Figure 10.8 shows a T-type flip-flop. The circuit is designed so that it has the truth table

t_n $(=T)$ (before clocking)	t_{n+1} $(=Q)$ (after clocking)
0	Q_n (no change)
1	$\overline{Q}_n$ (changes state)

In other words, $T=0$ produces no change in the output due to clocking, whereas $T=1$ directs a change in state due to clocking, whatever its state before clocking might have been. This is called *toggling*, hence the name T-type flip-flop. It now takes *two* clock pulses to restore the original state; the T-type flip-flop can be used to divide the number of pulses by a factor of 2. Suppose $T=1$ and $Q=1$ before clocking, then $Q=0$ after the first clock pulse and $Q=1$ after the second clock pulse, so the output completes *one cycle* for every two cycles at the input.

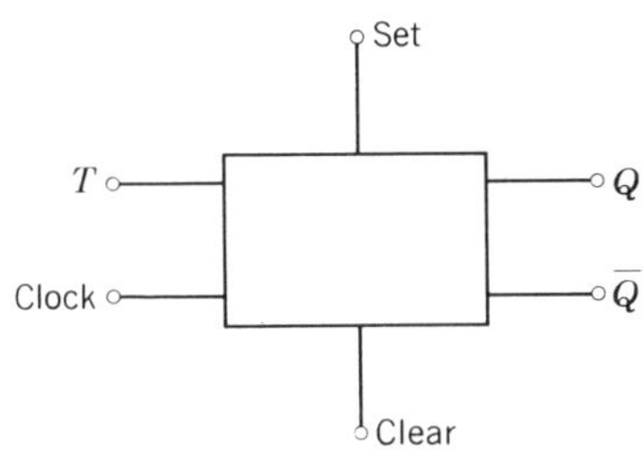

Figure 10.8. Symbol for T-type flip-flop.

10.2d The R-S Flip-Flop

The circuit in Fig. 10.9 shows one type of R-S flip-flop. It has the truth table

Input, t_n		Output, t_{n+1}
S	R	Q
0	0	Indeterminate
0	1	1
1	0	0
1	1	Q at t_n (no change)

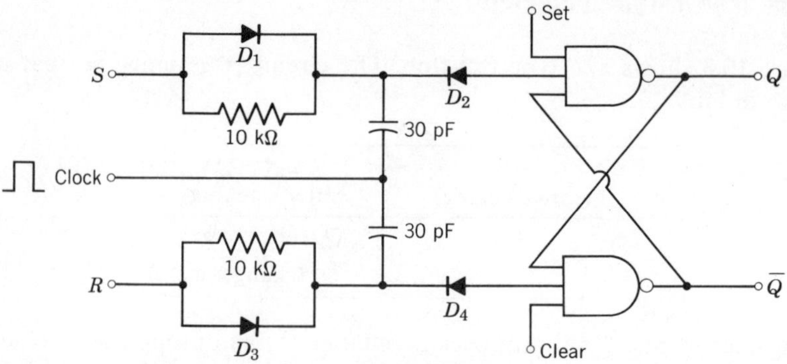

Figure 10.9. One type of R-S flip-flop. (From H. Malmstadt and G. Enke, *Digital Electronics for Scientists*, Copyright © 1969, W. A. Benjamin, Inc., Menlo Park, Cal.)

By using R as the input and feeding S from R through an inverter, the circuit becomes a form of the data latch discussed previously.

The flip-flop in Fig. 10.10 has the truth table

Input (before clocking)		Output (after clocking)
S	R	Q
0	0	Q_n (no change)
0	1	0
1	0	1
1	1	Indeterminate

That is, for $R=S=0$, Q does not change (the circuit remembers the last output). For $R \neq S$, Q is set to a 1 or a 0, independent of the previous state. This compares fully with Fig. 10.5.

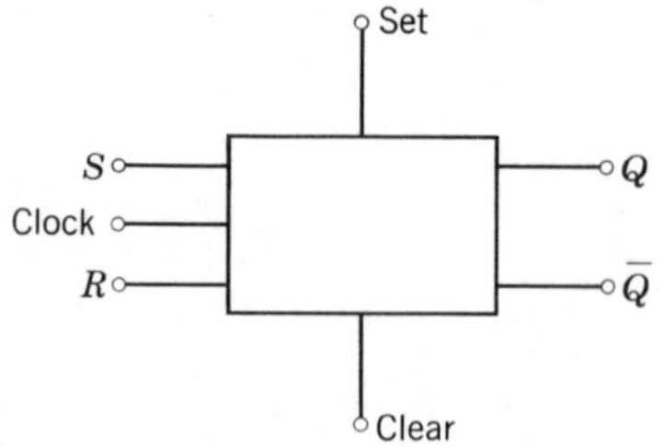

Figure 10.10. Symbol for another R-S flip-flop comparable to Fig. 10.5.

If set-clear terminals are present, the output for $R=S=1$ is determined by the inputs of these terminals, for example, clear$=0$, $Q=0$; set$=0$, $Q=1$.

When the Q output is in the logical 0 state before clocking, and $S=0$, it does not matter whether R is 0 or 1; Q will always be in the logical 0 state. When the Q output is in the logical 1 state before clocking, and $R=0$, it does not matter whether S is 0 or 1; after clocking Q will always be in the logical 1 state. We can also write the following truth table

Q_n	Q_{n+1}	R	S
0	0	Don't care[a]	0
0	1	0	1
1	0	1	0
1	1	0	Don't care[a]

[a] "Don't care" means that the outputs are independent of the value of R in line 1, and independent of the value of S in line 4.

This truth table shows when a change in state occurs. No change in state occurs for $Q_n=0$, $S=0$ or $Q_n=1$, $R=0$, whereas a change in state occurs for $Q_n=0$, $R=0$ or $Q_n=1$, $S=0$.

If more than one S or R inputs are present, the previous truth table remains the same, but S must be replaced by the Boolean product of all the S's, and R by the Boolean product of all the R's. For example, for three inputs S_1, S_2, S_3 and R_1, R_2, R_3, $S=S_1 \cdot S_2 \cdot S_3$ and $R=R_1 \cdot R_2 \cdot R_3$.

Figure 10.11*a* shows an *R-S* master-slave flip-flop. The clock input is normally 0, which keeps the outputs of gates 7 and 8 at 1 and prevents the R and S inputs from having any effect on the circuit. With the clock input at 0, the output of gate 9 is 1, which keeps gates 3 and 4 open. As a consequence the flip-flop of gates 1 and 2 will be in the same state as the flip-flop of gates 5 and 6. With the clock input at 1, the output of gate 9 is 0, which closes gates 3 and 4 and isolates the slave flip-flop from the master flip-flop. Gate 9 is a nand gate with the inputs tied together.

The current flowing out of gate 9 passes through a resistor R_0 to the clock input. Therefore the input voltage of gate 9 is somewhat higher than the zero state of the clock input. As a consequence gate 9 is turned on *before* gates 7 and 8 are turned on and is turned off *after* gates 7 and 8 are turned off. This is important for the timing, as is seen from Fig. 10.11 as follows. At time 1 of the clock pulse gates 3 and 4 close, isolating the slave flip-flop from the master flip-flop. At time 2 of the clock pulse gates 7 and 8 open, connecting the master flip-flop to the inputs. At time 3 gates 7 and

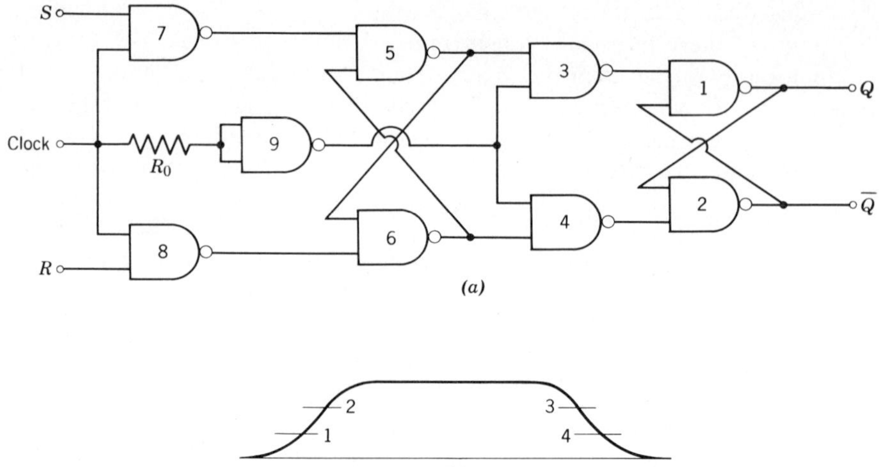

Figure 10.11. (*a*) *R-S* master-slave flip-flop. (*b*) Timing of flip-flop. (From H. Malmstadt and C. G. Enke, *Digital Electronics for Scientists*, Copyright © 1969, W. A. Benjamin, Inc., Menlo Park, Cal.)

8 close, isolating the master flip-flop from the inputs. At time 4 gates 3 and 4 open, connecting the master flip-flop to the slave flip-flop, thus permitting the latter to function.

The reader should verify that this circuit has the same truth table as Fig. 10.10.

Figure 10.12 is a MOS master-slave flip-flop schematic. It is included to indicate how larger MOS circuits can be made without having a single resistor in the circuit. This considerably simplifies circuit design and circuit manufacturing; it also allows greater density of packaging of the circuits.

10.2e The *J-K* Flip-Flop

The *J-K* flip-flop has the configuration shown in Fig. 10.13 and the truth table

Input, t_n (before clocking)		Output, t_{n+1} (after clocking)
J	K	Q
0	0	Q at t_n (no change)
0	1	0
1	0	1
1	1	$\overline{Q}_n$ (changes state)

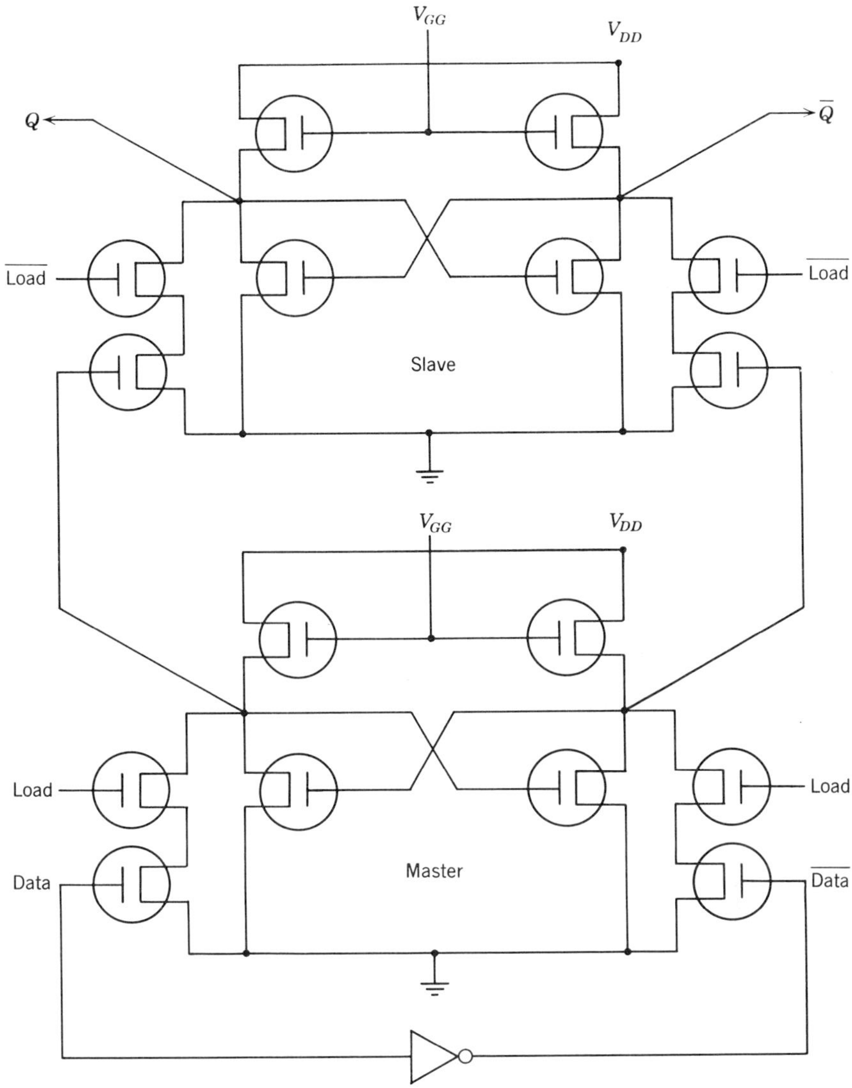

Figure 10.12. Example of MOSFET master-slave flip-flop. (From *MOS Integrated Circuits* edited by W. Penney and L. Lau © 1972 by Litton Educational Publishing, Inc. Reprinted by permission of Van Nostrand Reinhold Company.)

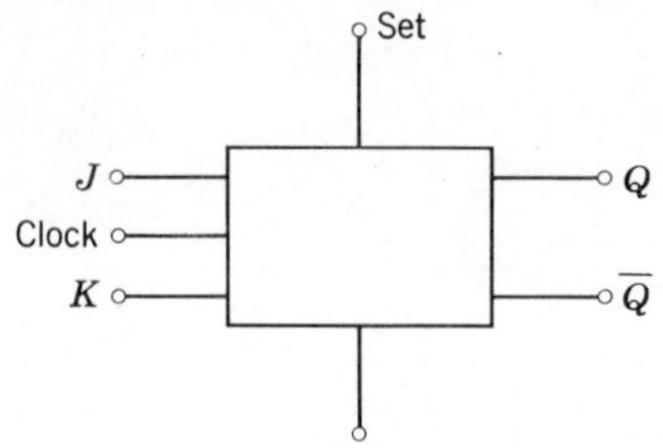

Figure 10.13. Symbol for J-K flip-flop.

Note that the (1, 1) state does not give an indeterminate output in this case. Note also that if J and K are tied together, one has the truth table corresponding to the T-type flip-flop in Fig. 10.8. By connecting K to J through an inverter, one obtains a D-type flip-flop.

$J=K=T$	Q_{n+1}
0	Q_n (no change)
1	$\overline{Q}_n$ (changes state)

We can also write the truth table as

Q_n	Q_{n+1}	J	K
0	0	0	Don't care[a]
0	1	1	Don't care[a]
1	0	Don't care[a]	1
1	1	Don't care[a]	0

[a] "Don't care" means that the outputs are independent of the value of K in lines 1 and 2 and independent of the value of J in the lines 3 and 4.

We can put this as follows. When $Q_n=0$, the J input controls the output Q_{n+1} after clocking, in that $Q_{n+1}=J$. When $Q_n=1$, the K input controls the output Q_{n+1} after clocking, in that $Q_{n+1}=\overline{K}$. No change in state occurs if $J=0$ and K is arbitrary when $Q_n=0$, or if $K=0$ and J is arbitrary when $Q_n=1$. A change $0\rightarrow 1$ occurs for $J=1$ and arbitrary K; a change $1\rightarrow 0$ occurs for $K=1$ and arbitrary J. The J-K flip-flop can be used for storage of information as in shift registers.

Figure 10.14 shows a J-K master-slave flip-flop. The right timing sequence is again determined by the resistor R_0 in series with gate 9, as in the previous case. The feedback operation from the output of gate 1 to one of the inputs of gate 8 and from the output of gate 2 to one of the inputs of

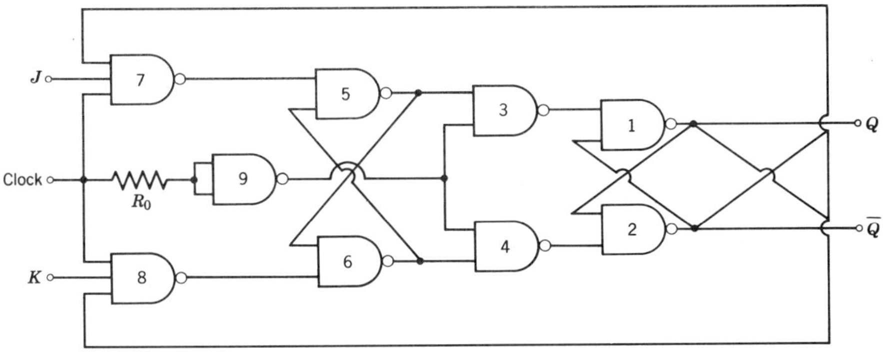

Figure 10.14. *J-K* master-slave flip-flop. (From H. Malmstadt and C. G. Enke, *Digital Electronics for Scientists,* Copyright © 1969, W. A. Benjamin, Inc., Menlo Park, Cal.)

gate 7 is introduced to acquire a toggle operation. Again flip-flop 5-6 (master) drives flip-flop 1-2 (slave) in a timing sequence similar to the one in the *R-S* flip-flop circuit in Fig. 10.11.

10.3 APPLICATIONS

10.3a Shift Registers

Since the output Q_{n+1} of the *J-K* flip-flop follows *J* if $K=\bar{J}$, we can build a shift register by connecting several *J-K* flip-flops in series, driving them from the same clock, and feeding the *K*-input of the first *J-K* flip-flop from the *J*-input through an inverter to ensure that $K=\bar{J}$. If a digit, 0 or 1, is now fed into the *J* input, the clocking pulse transmits it to the input of the next flip-flop. The next clocking pulse feeds the digit into the input of the third flip-flop and feeds a second digit into the input of the second flip-flop. Each further clock pulse shifts the information by one flip-flop and allows the feeding of one more digit into the input of the second flip-flop. The output *Q* of each flip-flop can be read at a special output terminal, so an *n*-digit binary number can be stored and read by feeding in exactly *n* clock pulses. The clock pulses *shift* the information, and the output terminals register it, hence the name shift register. By applying another set of *n* clock pulses one can take the information out or shift it to another shift register.

10.3b Counters

A *T*-type flip-flop can be used for counting purposes. Figure 10.15 shows one example. Suppose the flip-flops have been cleared by an appropriate

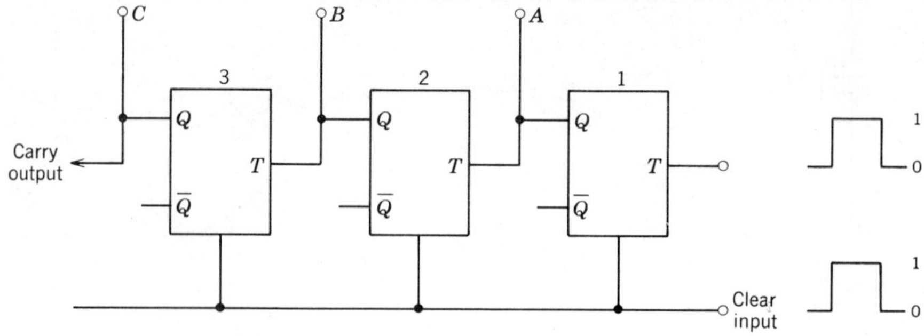

Figure 10.15. *T*-type flip-flop used as a counter.

pulse applied to the clear input, so that all outputs are 0s. If a *T*-input pulse comes along, output *A* becomes a 1. The next *T*-input pulse returns *A* to 0 and *B* to 1. The third *T*-input makes *A* equal to 1 and leaves *B* intact. In this manner one can read the number of counted pulses from *A*, *B*, *C*, ... in binary form.

With *N* flip-flops one can count $2^N - 1$ pulses.

PROBLEMS

10.1 In the circuit in Fig. 10.1, $C_0 = 30$ pF and $R_0 = 2000\ \Omega$. Find the duration of the trigger pulse at the diode input (time required to drop from maximum to 10%).

10.2 Show in detail that the circuit in Fig. 10.11*a* has the truth table associated with Fig. 10.10.

10.3 Show that the circuit in Fig. 10.14 has the truth table shown for Fig. 10.13.

10.4 How many *J-K* flip-flop units are needed to store the number 999 in binary form?

10.5 How many *T*-type flip-flop units are needed to count 999 pulses?

CHAPTER

11 Other Technologies

In this chapter we discuss various MOS and bipolar transistor circuits which allow greater density of packaging by removing as many resistors from the design as possible, or by making the individual devices smaller and closer together by having the devices share common regions (merging).

Section 11.1 discusses the advantages of silicon-on-sapphire MOS technology. Section 11.2 considers integrated injection logic (I^2L), base-coupled transistor logic (BCTL), complementary transistor logic (CTL), and I^2L with Schottky barrier diodes.

Some of these circuits are examples of merged transistor logic (MTL) in which individual transistors share one or more regions; this merging principle can be applied to both MOS structures and bipolar transistor circuits.

11.1 SILICON-ON-SAPPHIRE TECHNOLOGY

It is possible to grow thin films (0.5—1 μm thick) of single-crystal silicon of reasonable quality on a sapphire substrate. MOS circuits can be built with the help of such layers. This technology has the following advantages:

1. There is no substrate, hence no capacitances between the source and the substrate and between the drain and the substrate, nor influencing of the conducting channel by the substrate. This speeds up the response.

2. True isolation of individual devices is achieved, because the silicon is etched away where it is not needed.

3. The process is compatible with the CMOS process and with low-voltage CMOS technology.

4. No field inversion due to thick-film wiring occurs, because there is no silicon between the devices. In MOS devices made on a silicon chip the voltage in the wiring interconnecting the devices may induce a conducting channel between these devices, thus causing interdevice leakage. To prevent this, one puts *channel stoppers* between the devices, which interrupt possible conducting channels and so prevent leakage (a channel stopper is a highly conducting region of the same conductivity type as the substrate).

5. The devices are smaller, hence denser packaging is possible.

11.2 OTHER CIRCUITS

11.2a Integrated Injection Logic (I^2L)

Figure 11.1*a* shows an I^2L circuit in which the base current of an *n-p-n* transistor Q_2 is provided by injecting current into the base via a *p-n-p* transistor Q_1; this eliminates the base resistance for Q_2.

The implementation of such a circuit is shown in Fig. 11.1*b*; it consists of a lateral *p-n-p* transistor (Q_1) and a vertical *n-p-n* transistor (Q_2). The base of Q_1 is merged (i.e., is in common) with the emitter of Q_2, and the collector of Q_1 is merged with the base of Q_2. Q_1 acts as a current generator feeding the base of Q_2. The buried layer is introduced to decrease series resistance effects.

Figure 11.1*c* shows the equivalent circuit of Q_2 with its device capacitances C_E, C_C, and C_S, while Q_1 is represented by a current generator I_0.

It should be noted that Q_1 is used only as a current generator; the speed of the circuit is determined by the transistor Q_2. This is fortunate, for it is easier to make a high-speed vertical transistor than a high-speed lateral transistor.

The voltage applied to the devices is low; V_{EB1} is about 0.70 V. If Q_1 acts as a current generator feeding into Q_2, V_{EC1} is very small, hence V_{BE2} is large enough to turn Q_2 on. Because of the current gain of Q_2, the collector of Q_2 has a larger current-carrying capability than Q_1.

The circuit operates as follows. If the input voltage is high, the current-carrying capability of the collector of Q_2 is larger than the current-carrying capability of the *p-n-p* transistor Q_3 of the next stage. As a consequence, the collector voltage adjusts itself to the low value for which the currents are equal, and this turns off the *n-p-n* transistor Q_4 of the next stage. If the input voltage is low, the collector of Q_2 is turned off, and the *p-n-p* current

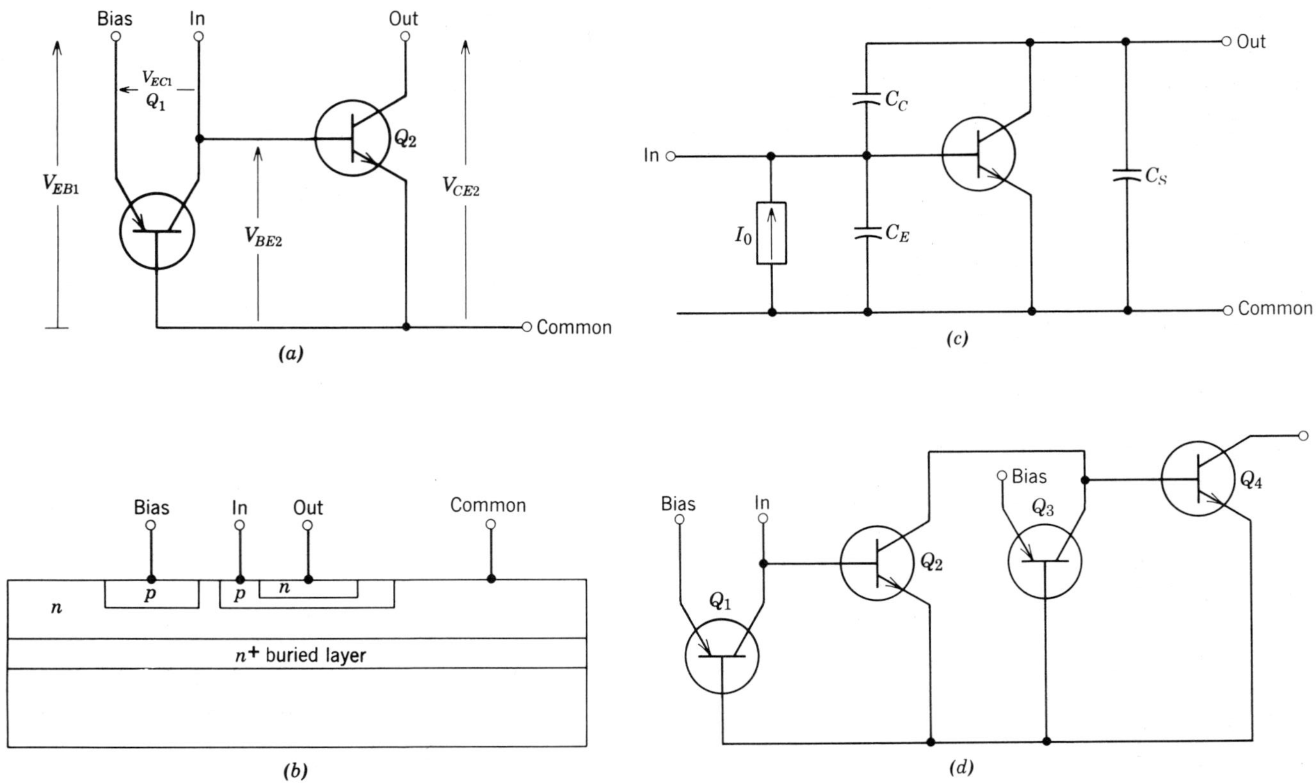

Figure 11.1. (*a*) I^2L circuit. (*b*) Layout of circuit on chip. (*c*) Equivalent circuit. (*d*) Explanation of the turn-on and turn-off of the stage.

generator of the next stage raises the base voltage of Q_4 to a sufficiently high level to turn it on.

The advantage of this technology is that the devices can be easily manufactured, occupy less space (since resistors are eliminated), and operate at low voltages. It is easy to provide the device Q_2 with more than one collector; as we shall see, this opens up new circuit possibilities. Finally, relatively low voltages can be used.

Figure 11.2*a* shows how a flip-flop can be made in this manner. Q_3 and Q_4 (*p-n-p*) act as current generators, Q_1 and Q_2 (*n-p-n*) form the flip-flop. The connected electrodes are merged together on a single chip; the layout on the chip is shown in Fig. 11.2*b*. The vertical rectangle is *p*-type, the bulk is *n*-type, and the horizontal rectangles are *p*-type except for the *n*-type squares that form the collectors. Figure 11.2*b* also shows how the flip-flop connections are made. No resistors are needed; only a current I_S is required, which is split between the two halves of the circuit. This is of course a low-voltage circuit.

The advantages of this circuit are that no resistors are needed and that superfluous regions are eliminated by merging; this allows much denser packaging of the circuits.

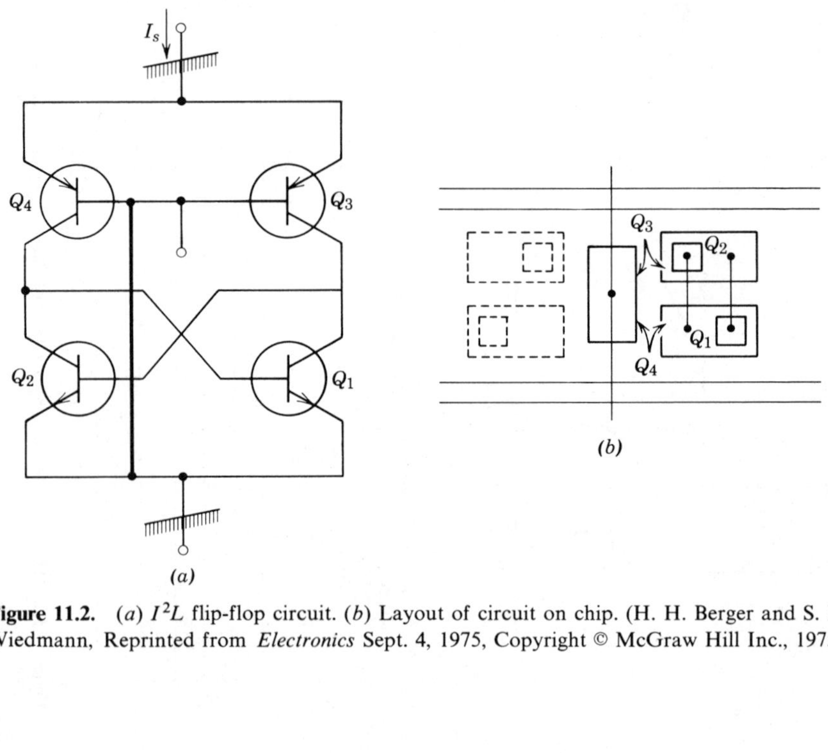

Figure 11.2. (*a*) I^2L flip-flop circuit. (*b*) Layout of circuit on chip. (H. H. Berger and S. K. Wiedmann, Reprinted from *Electronics* Sept. 4, 1975, Copyright © McGraw Hill Inc., 1975.)

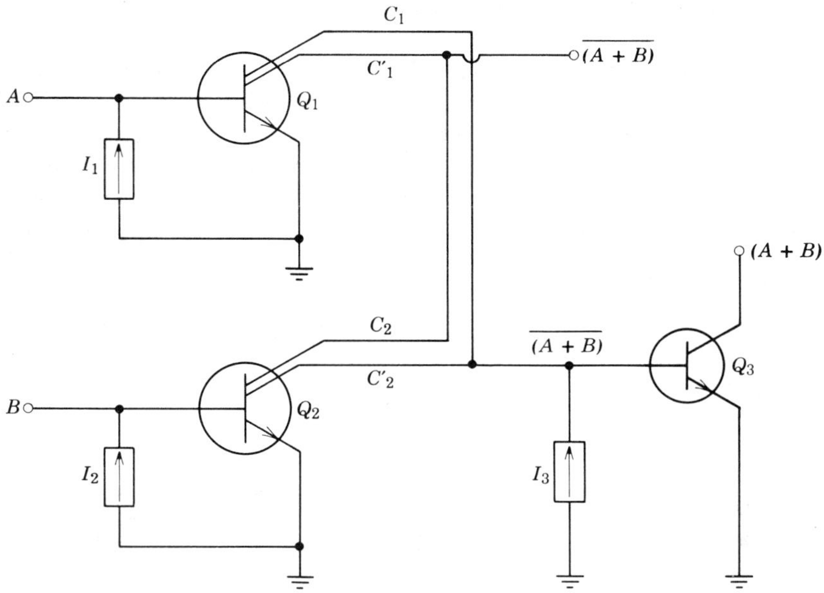

Figure 11.3. I^2L circuit with two collectors giving a nor and an or output. (R. L. Horton, J. Englade, and G. M. McGee, Reprinted from *Electronics,* Febr. 6, 1975, Copyright © McGraw Hill Inc., 1975.)

To demonstrate what can be done when an I^2L circuit is provided with more than one collector, we turn to Fig. 11.3. It shows a gate with two inputs and with a nor and an or output. To understand the operation of the circuit we first turn to the or output. Since it is the output of a normal inverter, the base of Q_3 represents the nor function $\overline{A+B}$; that is, the base voltage of Q_3 is low unless $A=B=0$. This is easily understood, since if $A=B=0$, then Q_1 and Q_2 are off, so the current generator I_3 flows into the base of Q_3, which makes the base about 0.70 V (1 output). If A is a 1, the base of Q_1 is high, thus turning the collector of Q_1 on. This allows for the flow of I_3 into the collector of Q_1, thereby lowering the base voltage of Q_3 to a low value (0 output). The same occurs if B is a 1, or if $A=B=1$. Hence the base of Q_3 represents $\overline{A+B}$, so that the output of Q_3 represents $A+B$.

We next turn to the nor output. The circuit shown is not complete; the terminal has to be fed into a similar I^2L circuit. When this is done, the base current generator of that circuit has the same function as in the other channel; it makes the circuit into a nor output.

11.2b Base-Coupled Transistor Logic (BCTL)

The basic element is a base-coupled current-mode switch. Figure 11.4*a* shows an or-nor gate, and Fig. 11.4*b* an and-nand gate based on this principle. Since the circuits operate in a current-switching mode (they switch emitter current rather than base current), the circuits are extremely fast. The logical 1 output is at 0 V, and the logical 0 output is at −0.60 V. While there are resistors in the circuit, they have relatively low values, so

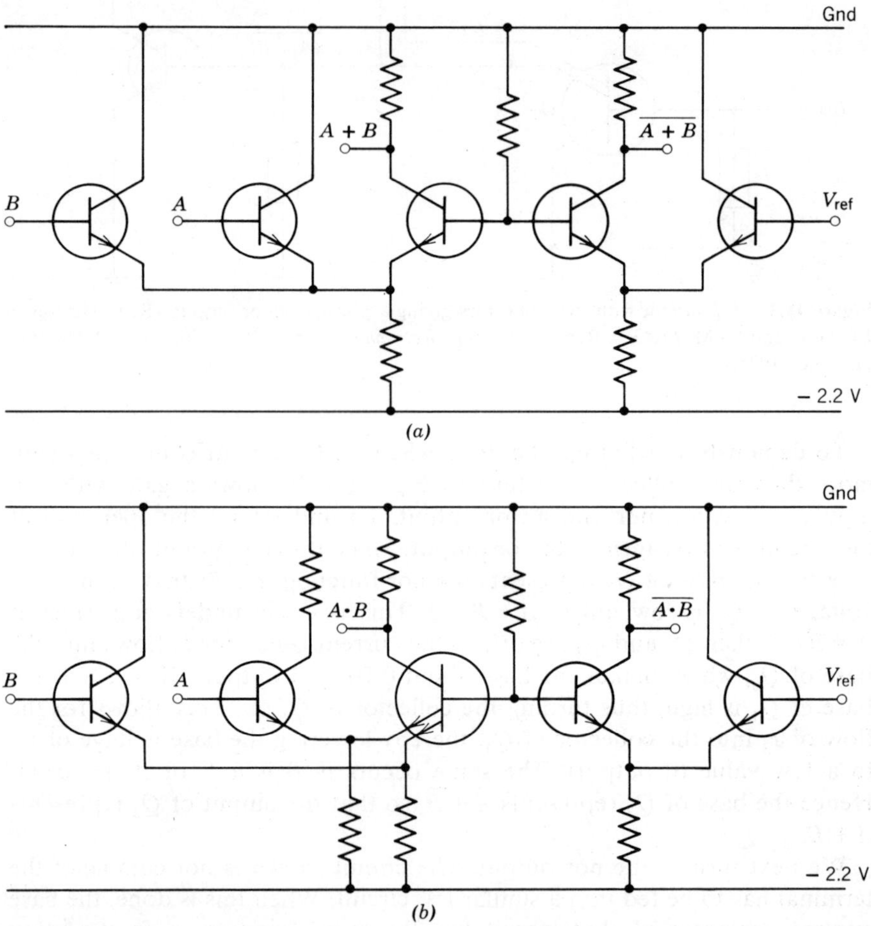

Figure 11.4. BCTL circuits. (*a*) Or-nor gate. (*b*) And-nand gate. (J. H. Gilder, *Electronic Design*, **5**, March 1, 1976, Copyright © Hayden Publishing Co. Inc., Rochelle Park N. J. 1976, by permission.)

the circuits can be densely packed. As the example shows, the total voltage involved is relatively small.

11.2c Complementary Transistor Logic (CTL)

The CTL circuit uses principles similar to those of the CMOS circuit. Figure 11.5 shows the configuration; the bases are in common and the collector of the *p-n-p* transistor is connected to the collector of the *n-p-n* transistor. One of the transistors is normally off, and the other is on; a 1 input produces a 0 output because Q_2 is on, and a 0 input produces a 1 output because Q_1 is on. If the *n-p-n* transistor is on (V_{BB} high), $V_0 \cong 0\,V$, since the collector current of that transistor discharges the load capacitance C_0 until its voltage reaches a value of about 0 V. If the *p-n-p* transistor is on (V_{BB} low), $V_0 \cong V_{CC}$, since the collector current of that transistor charges the load capacitance C_0 until the voltage reaches a value nearly equal to V_{CC}. Because the base-emitter voltage of either transistor should not exceed 0.70 V, V_{BB} and V_{CC} should not exceed 0.70 V. The circuit is therefore a low-voltage circuit.

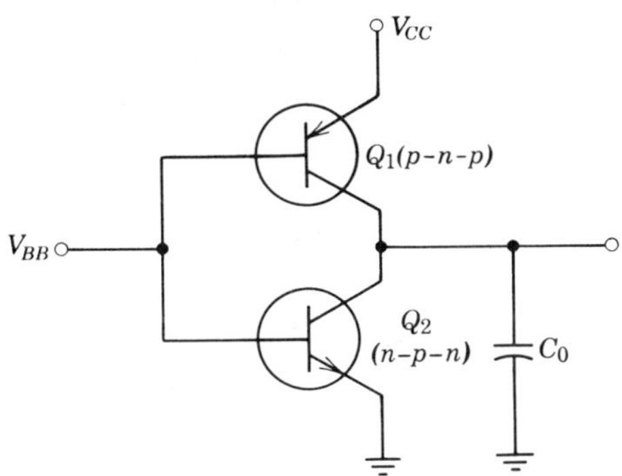

Figure 11.5. CTL circuit.

11.2d I²L with Schottky Barrier Output Diodes

It is easy to combine I²L with Schottky barrier diode technology. Figure 11.6*a* shows an I²L circuit with three Schottky barrier output diodes. Figure 11.6*b* shows its practical implementation. Figure 11.6*c* shows an I²L circuit with three input Schottky barrier diodes coming from previous stages.

The advantage of these circuits is that the Schottky barrier diode has a lower turn-on voltage, so the magnitude of the logic swing is reduced.

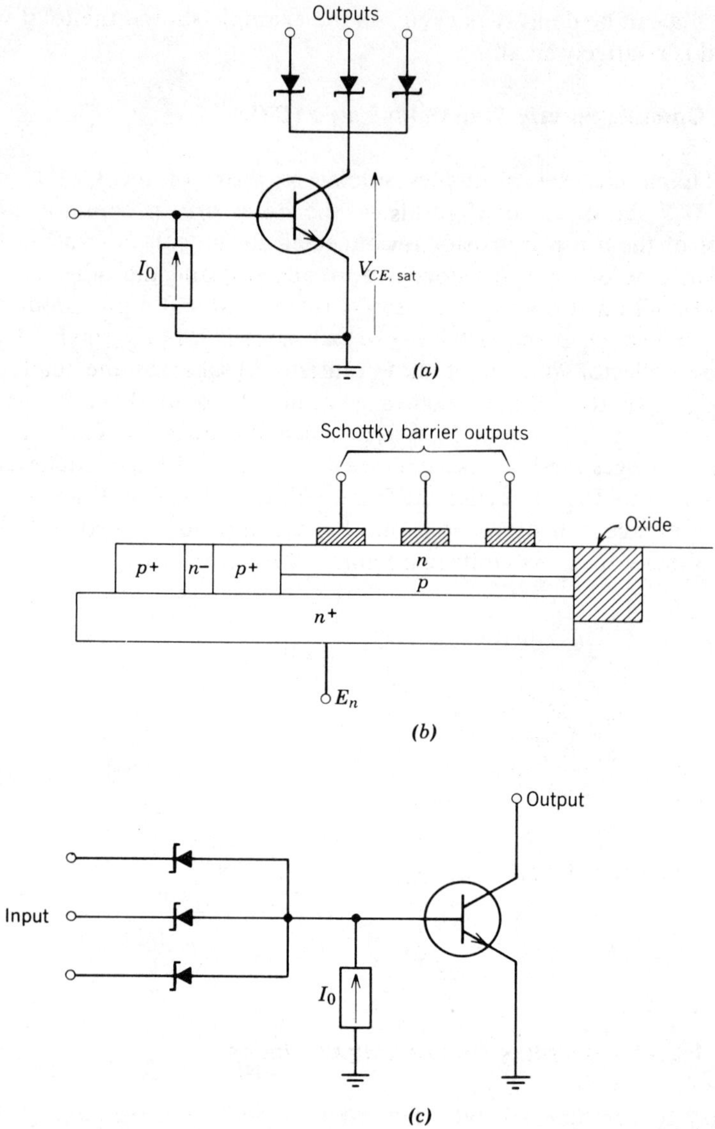

Figure 11.6. I^2L with Schottky barrier output diodes. (*a*) Circuit with three Schottky barrier output diodes. (*b*) Layout of circuit on chip. (*c*) Circuit with three Schottky barrier input diodes coming from previous stages.

11.2e Merged Logic (MTL)

Any logic device in which one or several transistors share regions (merged regions) involves *merged logic*. I^2L is a good example of merged logic, but it is not the *only* example, for merging applies not only to bipolar transistors but also to MOSFETs and JFETs.

The full-scale application of this principle allows many new configurations. We refer the reader to a review paper by Warner for some of the possibilities.*

* R. M. Warner, Jr., I Squared L, A Happy Merger, *IEEE Spectrum*, 42–76 (May 1976).

APPENDIX

1 The Equivalent Circuit of a Lossless Transformer

If a transformer has an input inductance L_1, an output inductance L_2, and a coupling inductance M, and if a source emf v_s with a internal resistance R_s is connected to the input and a load resistance R_L connected to the output (Fig. A.1*a*), one has the equations

$$i_1(R_s+j\omega L_1)-i_2 j\omega M=v_s$$

$$-i_1 j\omega M+i_2(R_L+j\omega L_2)=0 \tag{1}$$

Solving for i_2 yields

$$i_2=v_s\frac{j\omega M}{(R_s+j\omega L_1)(R_L+j\omega L_2)+\omega^2M^2} \tag{2}$$

$$v_0=i_2R_L=\frac{M}{L_1}v_s\frac{R_L}{\dfrac{L_2}{L_1}R_s+R_sR_L/j\omega L_1+R_L+j\omega L_2(1-M^2/L_1L_2)} \tag{3}$$

as is found by bringing $j\omega L_1$ outside the parentheses in the denominator. Substituting the turns ratio $n=(L_2/L_1)^{1/2}$ and the coupling factor $k=\dfrac{M}{(L_1L_2)^{1/2}}$, this expression can be written

$$v_0=knv_s\frac{R_L}{R_L+n^2R_s+n^2R_sR_L/(j\omega L_2)+j\omega L_2(1-k^2)} \tag{3a}$$

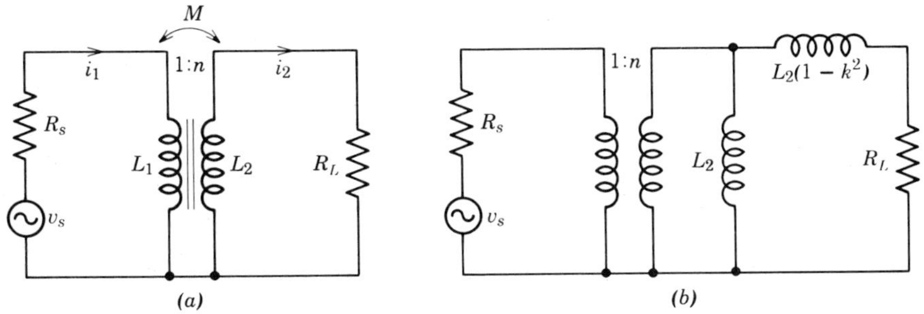

Figure A.1. (*a*) Ideal transformer connecting a source to a load. (*b*) Equivalent circuit.

For a good transformer k is very close to unity (e.g., 0.9995) so $L_2(1-k^2) \ll L_2$. No significant error is made by replacing k by unity in the term knv_s.

We now turn to the equivalent circuit in Fig. A.1*b*. It yields, if we bear in mind that $L_2(1-k^2) \ll L_2$,

$$v_0 = nv_s \frac{R_L}{R_L + n^2R_s + n^2R_sR_L/(j\omega L_2) + j\omega L_2(1-k^2)} \tag{4}$$

which corresponds to Eq. (3a) if k is very close to unity.

The transformer can thus be represented accurately by an ideal transformer of a step-up ratio n with a parallel inductance L_2 and a spreading inductance $L_s = L_2(1-k^2)$;it is used in Fig. 8.4*b*.

APPENDIX

2 General Discussion of Enhancement Mode MOSFET Loads

In Chapter 4 we discussed enhancement-mode circuits in which one of the devices, Q_2, was used as a load and the gate of Q_2 was connected to the supply voltage V_{DD}, so Q_2 was always saturated.

We here discuss the more general case in which Q_1 and Q_2 are n-channel enhancement-mode devices, the turn-on voltages V_{T1} (>0) and V_{T2} (>0) are different, and the gate of Q_2 is connected to a separate supply voltage V_{GG} (Fig. A.2). We then have the following equations.

For Q_1 the characteristic is

$$I_{D1} = K_1\left[2(V_i - V_{T1})V_0 - V_0^2\right] \qquad \text{for } V_0 < V_i - V_{T1} \tag{5}$$

(Q_1 not saturated), and

$$I_{D1} = K_1(V_i - V_{T1} - V_0)^2 \qquad \text{for } V_0 > V_i - V_{T1} \tag{6}$$

(Q_1 saturated). For Q_2 we have

$$I_{D2} = K_2\left[2(V_{GG} - V_{T2} - V_0)(V_{DD} - V_0) - (V_{DD} - V_0)^2\right] \tag{7}$$

for $V_{DD} - V_0 < V_{GG} - V_{T2} - V_0$ or $V_{DD} < V_{GG} - V_{T2}$ (Q_2 not saturated) and

$$I_{D2} = K_2(V_{GG} - V_{T2} - V_0)^2 \qquad \text{for } V_{DD} > V_{GG} - V_{T2} \tag{8}$$

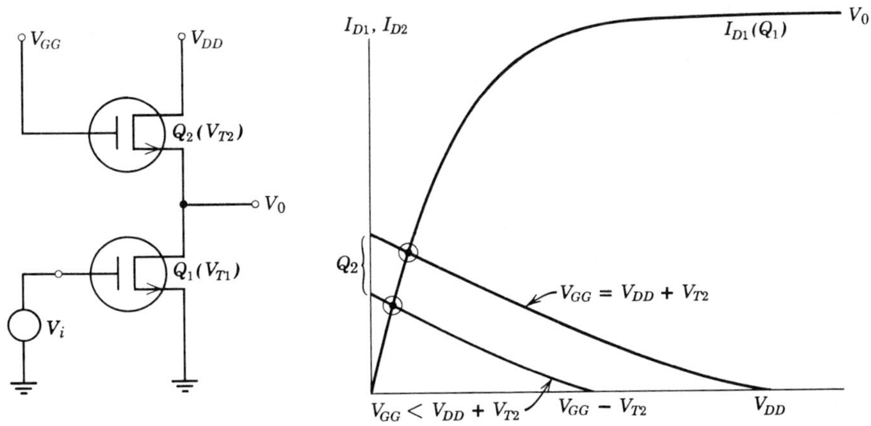

(a) (b)

Figure A.2. (*a*) Integrated inverter in which the gate of the load MOSFET is fed from a separate voltage. (*b*) Determination of the operating point.

(Q_2 saturated). We note that in the first case $I_{D2}=0$ for $V_0=V_{DD}$, whereas in the second case $I_{D2}=0$ for $V_0=V_{GG}-V_{T2}$. The operating point lies where

$$I_{D1}(V_0)=I_{D2}(V_0)=I_{\text{on}} \qquad \text{at } V_0=V_{\text{on}} \tag{9}$$

We observe that I_{on} and V_{on} decrease with decreasing V_{GG}, so one can adjust the operating point by the choice of V_{GG}. The current at $V_0 \cong 0$ reaches its highest value when $V_{GG}=V_{DD}+V_{T2}$ and decreases when V_{GG} is further increased. There is no particular merit in having $V_{GG}>V_{DD}+V_{T2}$.

APPENDIX

3 Positive and Negative Logic

Here we summarize the definitions of positive and negative logic. Positive logic means 1 is a high voltage and 0 is a low voltage; negative logic means 1 is a low voltage and 0 is a high voltage.

In many of our examples a 0 input was close to 0 V, a 1 input is then a large positive voltage for positive logic, and a large negative voltage for negative logic. But this is of course unnecessary, since the levels may be shifted at will, as long as their relative position does not change.

We thus have four possibilities, as illustrated in Fig. A.3. Figure A.3*a* shows the normal definition of positive logic with a 0 near 0 V. By interchanging 0 and 1 we obtain Fig. A.3*b*, which is a case of negative logic. Figure A.3*c* shows the normal definition of negative logic with a 0 near 0 V. By interchanging 0 and 1 we obtain Fig. A.3*d*, which is a case of positive logic.

We should bear in mind that by interchanging 0 and 1 we make the following transformations:

And becomes or

Or becomes and

Nand becomes nor

Nor becomes nand

An and circuit for positive logic is thus an or circuit for negative logic, and so on.

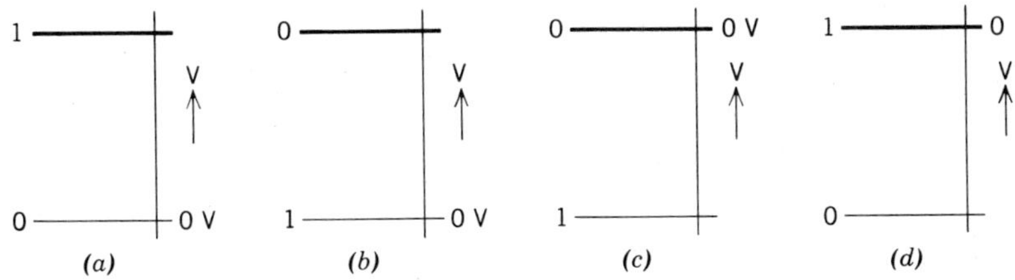

Figure A.3. (*a*) Normal definition of positive logic. (*b*) Negative logic made from (*a*) by inverting 0 and 1. (*c*) Normal definition of negative logic. (*d*) Positive logic made from (*c*) by inverting 0 and 1.

APPENDIX

4 Fan-Out Under Worst-Case Design

In Chapter 7 we discussed the fan-out for the case that the circuits themselves were operating under worst conditions. But device and circuit parameters can also vary from circuit to circuit, and this gives rise to further restrictions on the fan-out capability. These variations are:

1. *Variation in β_F.* Here the lowest possible value of β_F should be used in worst-case design considerations.

2. *Variation in V_{CC}.* Often the power supply is not too well regulated. Variations of $\pm 10\%$ in V_{CC} are not uncommon; therefore the lowest value of V_{CC} must be used.

3. *Variation in resistance values.* Often the fan-out expressions involve only resistance *ratios*, and these vary little for different circuits of the same type; when that is true, this is not a problem.

4. V_{BE0} and $V_{CE,\text{sat}}$ can vary from unit to unit, and this should be taken into account in an appropriate manner.

If one allows too large a fan-out, some circuits may not operate properly; this adversely affects the yield of the integrated processing.

APPENDIX

5 Solutions to Selected Problems

1.1 (a) $5.0+10.0\exp(-t/\tau)$ for $0<t<T/2$.
(b) $5.0+7.31\exp(-t/\tau)$ for $0<t<T/2$.
(c) $5.0+5.25\exp(-t/\tau)$ for $0<t<T/2$.

For $T/2<t<T$: opposite sign and t replaced by $t-T/2$.

1.2 (a) $5.00\exp(-t/\tau)$ for $0<t<T/2$.
(b) $7.31\exp(-t/\tau)$ for $0<t<T/2$.
(c) $10.00\exp(-t/\tau)$ for $0<t<T/2$.

For $T/2<t<T$: opposite sign and t replaced by $t-T/2$.

1.3 (a) 33.3 μF.
(b) 1.0% per ms.

1.4 (a) 0.10 μF.
(b) 1.0% per ms.

1.5 (a) 0.693×10^{-7} s.
(b) 1.099×10^{-7} s.
(c) 1.792×10^{-7} s.
(d) 3.045×10^{-7} s.

1.6 (a) $V_0(t)=-5.0$ V for $t<0$.
(b) $V_0(t)=-25-10^{-3}t$ for $0<t<50$ ms.
(c) $V_0(t)=-5.0-10\exp(-t/\tau)$ for $t>50$ ms, where $\tau=2.5\times10^{-2}$ s.

1.7 (a) $V_0=20[1-\exp(-t/\tau_a)]$ for $0<t<t_1=50$ ms, $\tau_a=1$ ms.
(b) $V_0=11.53+5.77\exp[-(t-t_1)/\tau_b]$, $\tau_b=3.33$ μs, $t>t_1=50$ ms.

2.1 0.150 mA, 0.651 V.

2.2 (a) 0.77 Ω.
(b) −62 V.
(c) 42.3%.
(d) 8.2 A.
(e) 9.15 h.

2.3 (a) 344 W.
(b) 3.15 kWh.

2.4 4.30 V for 10 V peak to peak amplitude.

2.5 (a) Region I: $v_1 < 8$ V, $v_0 = 8.0$ V, $i_1 = 0$.
(b) Region II: $8 < v_1 < 15$ V, $v_0 = (2v_1 + 40)/7$ V, $i_1 = (5i_i - 40)/28$ mA.
(c) Region III: $15 < v_1 < 30$ V, $v_0 = \frac{2}{3}v_1$ V, $i_1 = v_1/12$ mA.
(d) Region IV: $v_1 > 30$ V, $v_0 = (2v_1 + 40)/5$ V, $i_1 = (3v_1 - 40)/20$ mA.

2.6 4.30 V.

2.7 3.23 V.

2.8 (a) $V_0 = -5 + 10 \cos\omega t$ V, $V_0 = 10 \cos\omega t$ V, $V_0 = 10 \cos\omega t$ V.

2.9 $0.683 > V_D > 0.563$ V.

2.10 3900 pF.

2.11 50 pF.

2.12 (a) 5300 Ω.
(b) 6.8×10^{-9} s.

2.13 (a) 0.58×10^{-7} s.
(b) 2.15×10^{-7} s.

2.14 (a) 2.5×10^{-9} s.
(b) 2.18×10^{-8} s.

2.15 $V_D(t) = 25.8 \times 10^{-3} \ \ln[\exp(27.1 - t/\tau) + 1] \ \text{V} \cong 25.8 \times 10^{-3} (27.1 - t/\tau)$ V for $t/\tau < 25$.

2.16 (a) 20 pC.
(b) 2.47×10^{-8} s.
(c) 24.7 pC.

3.2 0.760 V.

3.3 0.718 V.

3.4 (a) $I_B/I_C = 0.01$: $V_{CE} = \infty$, $I_B/I_C = 0.02$: $V_{CE} = 0.176$ V, $I_B/I_C = 0.05$: $V_{CE} = 0.141$ V, $I_B/I_C = 0.10$: $V_{CE} = 0.121$ V, $I_B/I_C = 0.20$: $V_{CE} = 0.105$ V, $I_B/I_C = 0.50$: $V_{CE} = 0.086$ V.
(b) $I_B/I_C = 1.39 \times 10^{-2}$.

3.5 1.09 V.

3.6 (a) $V_E=(V_{BB}-V_{BE0})R_e(1+\beta_F)/[R_b+R_e(1+\beta_F)]$.
(b) $V_E=[(V_{BB}-V_{BE0})/R_b+(V_{CC}-V_{CE,\text{sat}})/R_c]/(1/R_e+1/R_b+1/R_c)$.
(c) $V_E\cong(V_{CC}-V_{CE,\text{sat}})/(1+R_c/R_e)$, $V_{BB}\cong V_{BE0}+V_E$.

3.8 (a) 2.48×10^{-9}s.
(b) 2.42×10^{-9} s.
(c) 11.5×10^{-8} s.

3.9 2.3×10^{-8} s versus 11.5×10^{-8} s.

3.10 $t_1=2.28\times10^{-9}$ s, $t_2=11.5\times10^{-8}$ s.

3.11 6.0 V.

3.12 V_c varies from 6.15 to 5.85 V.

3.13 3.86 V.

3.14 From 3.86+0.065 V to 3.86−0.065 V.

3.15 4.15 V.

3.16 τ_π varies from 3.9×10^{-8} s at $\beta_F=100$ to 1.16×10^{-8} s for $\beta_F=30$, τ_T varies from 19×10^{-8} s to 5.7×10^{-8} s.

3.17 50×10^{-12} pF.

3.18 3.3×10^{-8} s.

3.19 $t_0=7.0\times10^{-9}$ s, $t_1=1.00\times10^{-8}$ s, $t_3=1.25\times10^{-8}$ s, $t_2=0.70\times10^{-8}$ s, total turn-on time 1.70×10^{-8} s, total turn-off time 1.95×10^{-8} s.

3.20 $t_0=1.0\times10^{-9}$ s, $t_1=1.00\times10^{-8}$ s, $t_3=5.3\times10^{-8}$ s, $t_2=4.8\times10^{-8}$ s, total turn-on time 1.1×10^{-8} s, total turn-off time 10.1×10^{-8} s.

3.21 (a) $R_c=1185\ \Omega$, $R_b=7200\ \Omega$, $R_e=1130\ \Omega$.
(b) Rise time 1.08×10^{-8} s (90% response), fall time 7.5×10^{-8} s.
(d) $t_3=0.34\times10^{-8}$ s.

4.1 (a) Yes, $R_{\text{on}}=167\ \Omega$.
(c) approximate 0.67 V, accurate 0.73 V.

4.2 (a) Yes.
(b) $R_{\text{on}}=167\ \Omega$.
(c) $V_0\cong-0.67$ V, accurate −0.73 V.

4.3 (a) No, $V_0=6.0$ V.

4.4 Device is off, $V_0=10.0$ V.

4.5 (a) Yes.
(b) −1.30 V accurate, −1.25 V approximate.

4.6 Yes.

4.7 (a) $V_0=-17.0$ V, $V_i>-3.00$ V.
(b) $V_0=-3.28$ V approximate, −2.87 V accurate.

4.8 (a) $C_T = 27.0 \times 10^{-12}$ F.
(b) 6.75×10^{-8} s.

4.9 7.9×10^{-9} s.

4.10 2.3×10^{-7} s.

4.11 (a) $I_{D\max} = 36$ mA.
(b) $V_{p0} = 3.0$ V.
(c) 5.0×10^{-9} s.
(d) The same.
(e) 4.0×10^{-9} s.

5.1 V_{02} switches from 0.10 V to 2.90 V, V_{01} from 2.90 V to 0.10 V.

5.2 V_{02} switches from 1.8 V to 20 V, V_{01} from 20 V to 1.8 V.

5.3 V_{02} switches from -2.21 V to -16.00 V, V_{01} from -16.00 V to -2.21 V.

5.4 Factor 5.

5.5 (a) $C_1 = 5.0$ pF.
(b) Factor 3.

5.6 (a) 126 pF.
(b) 290 V.
(c) -4.00 V on 4.00 V dc background.

5.7 (a) 171 pF.
(b) 4.23 V, -6.00 V on 6.00 V dc background.

5.8 (a) 580 pF.
(b) -19 V on -1.0 V dc background.
(c) $+19$ V on -20 V dc background.

5.9 6300 pF.

5.10 $V_{CE2}(t)$ has 1.82 V pulse on 0.20 V background, $V_{CE1}(t)$ has -3.80 V pulse on 4.0 V background, $V_{BE2}(t)$ has -3.80 V triangular pulse on 0.70 V background, $V_{BE1}(t)$ has 0.50 pulse on 0.20 V background.

5.11 1980 pF.

5.12 $V_{1H} = 3.90$ V, $V_{1L} = 3.0$ V; $V_{0L} = 8.65$ V, $V_{0H} = 10.0$ V, transistors are not saturated.

7.1 $I_B = 1.500$ mA, $I_C = 2.700$ mA, $I'_B = 0.090$ mA, $I'_C = 4.50$ mA.

7.2 (a) $V_A = 0.70$ V.
(b) $V_B = 0.90$ V.

7.3 $V_A = -3.0$ V.
(b) $V_B = -4.0$ V.

7.4 0.90 V.

7.5 1.56 V.

7.6 100.

7.7 (a) $I_{BS} = 0.10$ mA.
(b) $I_A = 1.00$ mA.

7.8 (a) Three diodes.
(b) 4.10 V.

7.9 13.

7.10 (a) $t_1 = 7.0 \times 10^{-9}$ s, $I_{max} = 195$ mA, $t_d = 0.70 \times 10^{-10}$ s.

7.11 (a) 41.25 mA.
(b) 57.5 mA.
(c) 3.6×10^{-9} s.
(d) 4.5×10^{-9} s.

7.12 (a) 3.80 mA.
(b) 380 mA.
(c) 2.6×10^{-10} s.
(d) 101 mA.
(e) 1.73×10^{-9} s.

7.13 (a) $V_A = V_B \leqslant -1.50$ V.
(b) $V_4 = -1.56$ V, $V_5 = -0.70$ V.
(c) $V_A = V_B > -1.30$ V.
(d) $V_4 = -0.70$ V, $V_5 = -1.56$ V.

7.14 (a) $I_0 = 494$ mA.
(b) $I_1 = 18.9$ mA.
(c) Turn-on time 0.37×10^{-10} s, turn-off time 6.1×10^{-10} s.

7.15 Impedance level about a factor of 10 higher.

8.1 (a) 40,000 pF.
(b) 1075 Ω.

8.2 (a) 10 V.
(b) 0.84 μF.

8.3 (a) 318 pF, 2.0 A.

8.4 (a) As in the text, replace β_F by α_F.
(b) 4.1×10^{-5} s.

9.7 (b) 0.504×10^8 Hz.
(c) 0.15 V.

9.8 42.86 Ω.

9.10 $\omega CR = 1$, $\mu = 3$.

9.11 The same.

10.1 0.138 μs.

10.4 10 units.

10.5 10 units.

Bibliography

1. D. Casanent, *Digital Electronics*, Quantum Publishers, New York, 1974.
2. V. H. Grinich and H. G. Jackson, *Introduction to Integrated Circuits*, McGraw-Hill, New York, 1975.
3. D. A. Hodges (Ed.), *Semiconductor Memories*, The Institute of Electrical and Electronic Engineers, New York, 1972.
4. T. Kohonen, *Digital Circuits and Devices*, Prentice Hall, Englewood Cliffs, N. J., 1972.
5. D. A. Lynn, C. S. Meyer, and D. J. Hamilton, *Analysis and Design of Integrated Circuits*, McGraw-Hill, New York, 1967.
6. J. Millman and C. C. Halkias, *Integrated Electronics, Analog and Digital Circuits and Systems*, McGraw-Hill, New York, N. Y., 1972.
7. J. D. Ryder, *Electronic Fundamentals and Applications*, 5th Ed., Prentice Hall, Englewood Cliffs, N. J., 1976.
8. L. Strauss, *Wave Generation and Shaping*, 2nd Ed., McGraw-Hill, New York, 1970.
9. A. Barna and D. I. Porat, *Integrated Circuits in Digital Electronics*, Wiley, New York, 1973.
10. A. Van der Ziel, *Introductory Electronics*, Prentice Hall, Englewood Cliffs, N.J., 1974.

Index